THE IMPACT
OF SPACE SCIENCE
ON MANKIND

NOBEL SYMPOSIUM COMMITTEE (1975)

STIG RAMEL, *Chairman* • Executive Director, Nobel
Foundation

ARNE FREDGA • Chairman, Nobel Committee for
Chemistry

TIM GREVE • Director, Norwegian Nobel
Institute (Peace)

BENGT GUSTAFSSON • Secretary, Nobel Committee for
Medicine

LARS GYLLENSTEN • Member, Swedish Academy
(Literature)

LAMEK HULTHÉN • Chairman, Nobel Committee for
Physics

ERIK LUNDBERG • Chairman, Prize Committee for
Economic Sciences

NILS-ERIC SVENSSON • Executive Director, Bank of
Sweden Tercentenary Foundation

THE IMPACT
OF SPACE SCIENCE
ON MANKIND

Edited by
Tim Greve
Norwegian Nobel Institute

Finn Lied
*Norwegian Council for Scientific
and Industrial Research*

and
Erik Tandberg
Norconsult A.S.

PLENUM PRESS □ NEW YORK AND LONDON

Library of Congress Cataloging in Publication Data

Nobel Symposium, 31st, Spåtind, Norway, 1975.
   The impact of space science on mankind.

   Includes index.
   1. Scientific satellites—Congresses. 2. Astronautics in earth sciences—Congresses. 3.
Space sciences—Congresses. 4. Artificial satellites in telecommunication—Congresses.
I. Greve, Tim, 1926-   II. Lied, Finn. III. Tandberg, Erik, 1932-   IV. Title.
TL798.S3N6 1975   301.24'3   76-26652
ISBN 0-306-33701-0

Proceedings of the thirty-first Nobel Symposium on The Impact of
Space Science on Mankind held at Spåtind, Norway, September 7-12, 1975

ORGANIZING COMMITTEE

GREVE, TIM—*Director, Norwegian Nobel Institute*

INGVALDSEN, BERNT—*Director, Member of the Norwegian Nobel Committee*

LIED, FINN—*Chairman, Executive Board, Norwegian Council for Scientific and Industrial Research*

SANNESS, JOHN—*Director, Norwegian Institute for Foreign Affairs, member Norwegian Nobel Committee*

TANDBERG, ERIK—*Director, Norconsult A.S.*

EARLIER NOBEL SYMPOSIA

Symposia 1-17 and 20-22 were published by Almqvist & Wiksell, Stockholm and John Wiley & Sons, New York; Symposia 23-25 by Nobel Foundation, Stockholm and Academic Press, New York; Symposium 26 by the Norwegian Nobel Institute, Universitetsforlaget, Oslo; Symposium 27 by Nobel Foundation, Stockholm and Almqvist & Wiksell International, Stockholm; Symposium 28 to be published by Academic Press, New York; Symposium 29 by Nobel Foundation, Stockholm and Trycksaksservice AB, Stockholm; and Symposium 30 by Plenum Press, New York.

PREFACE

Nobel Symposium No 31 on The Impact of Space Science on Mankind was held at Spåtind, Norway, September 7 - 12, 1975. Twenty-
seven leading experts from the United States, the U.S.S.R. and
Western Europe attended the Symposium.

Four main subjects were discussed: The Impact of Space Science,
introduced by Professor Reimar Lüst; The Impact of Space Communica-
tion, introduced by Dr Joseph Charyk; The Impact of Earth Resources
Exploration from Space, introduced by Dr William Nordberg; and The
Impact of Space Assisted Meteorology, introduced by Dr Robert M
White.

This book contains edited summaries of the papers presented at
the Symposium and summaries of the discussions.

The Symposium was financed by the Nobel Foundation through
grants from the Tercentenary Foundation of the Bank of Sweden and
organized by a special committee appointed by the Norwegian Nobel
Institute.

Tim Greve
Finn Lied
Erik Tandberg

CONTENTS

ix

THE IMPACT OF SPACE SCIENCE*

R. Lüst

Max Planck Society for the Advancement of Science

1. NEW METHODS

Space science as a discipline is not so much.defined by the
objects to be investigated or observed as by the technical means –
namely balloons, sounding rockets, satellites and space probes – to
be applied for the investigation above the surface of the earth.
As in many other fields, scientific and technical progress are very
closely correlated, and space technology has provided the scientists
with completely new possibilities for their research of the earth,
its environment, the planetary system, astronomical objects, and the
universe.   These techniques can be applied in different ways:

A. Large parts of the surface of the earth and its atmosphere can
    be observed and monitored simultaneously.

B. The surroundings of the earth, the moon and the planets can be
    investigated directly by measuring instruments.

C. The electromagnetic and the corpuscular radiation approaching
    the earth can be observed from an extraterrestrial (observing)
    station not inhibited and influenced by the earth's atmosphere
    and by the earth's magnetic field.   Therefore the whole range
    of the electromagnetic spectrum as well as the energy spectrum
    of the arriving photons and particles can be exploited and
    yields new data on known objects as well as on objects so far

---

*The parts 2.5 – 2.9 have been written by Dr Klaus Beuermann
(Max-Planck-Institut für Physik und Astrophysik, Institut für
Extraterrestrische Physik, Garching bei München).   I want to
thank him also for many valuable discussions.   Furthermore I
am thankful to Dr Dieter Hovestadt for a contribution.

not yet detected.

D. Direct experiments can be carried out in the extraterrestrial
space under conditions not to be realized on the earth.

All these different ways have been used, and they have opened
the road for the application satellites. Thus it could easily be
proven that space science was the first step necessary to enable us
to use satellites for communication, to monitor the physical environment of man and to study its resources on the earth. In this way
the impact of space science on mankind can directly be demonstrated.
But also the impact on technical developments - particularly in
electronics - should be taken into account, e.g. micro-
miniaturization.

However, this is not the motivation for space science, and it
would be a serious mistake, if one would try to sell space science
on these reasons. Here we meet with the most difficult question:
why one should carry out basic research and why one is justified to
ask for funding basic research. From our experience of the past
we know that the results of basic research are the necessary corner
stones for our further technological and technical development.
But it is very difficult to predict which areas are essential for
the future development. But even this is not a sufficient motivation for supporting basic research, to which space science belongs.
The main motivation is to widen our knowledge about the nature,
which may have a profound impact on mankind as e.g. the discovery
of Copernicus has demonstrated, but which is very difficult to be
anticipated.

2. NEW DISCOVERIES

For these reasons this shall not be an attempt to describe in
detail what space science has accomplished so far. Instead, I
shall try to give a list of the findings and discoveries achieved
with the means of space research which could have a strong impact
on further scientific developments and thereby on mankind.

The progress in the understanding of the physics and the
behaviour of cosmical objects has been considerable, and also it
was possible to accumulate new knowledge in certain areas of physics
as e.g. in plasmaphysics. The following summary might serve as a
basis for discussion:

2.1 The Surrounding of the Earth. The *in situ* measurements in the
surrounding of the earth with the help of satellites and sounding
rockets have changed our picture of this region completely. The
space out to a distance of about 10 R_E in the direction towards the
sun is dominated by the earth's magnetic field. In the anti-solar

direction, the magnetic lines of force are not closed, but temporarily linked to the interplanetary field and stretched far beyond the lunar orbit to form a long magnetic tail. This whole region is called the magnetosphere.

The first important discovery by means of satellites was that of the radiation belts, which constitute the innermost part of the magnetosphere where the field is approximately dipolar. Most of the energetic charged particles trapped in the earth's magnetic field seem to originate from the solar wind and are accelerated by the variable electromagnetic field, whereby the energy is again derived from the solar wind.

The outer parts of the magnetosphere contain layers of low-energy plasma (average proton energies of a few 100 eV to a few keV). The interface between solar wind and magnetosphere, the magnetopause, is covered on the inside by a plasma layer which extends far into the tail. It is filled by ions and electrons which have just managed to penetrate through the barrier of the magnetopause. The greatest energy reservoir of the magnetosphere is the plasma sheet which occupies the central region of the tail and maps into the auroral oval. Particle and energy losses from the plasma sheet to the upper atmosphere are the origin of the auroral phenomena.

The various regions of the magnetosphere are subject to a large scale internal convection process which is initiated by momentum transfer from the solar wind and controlled by frictional forces acting in the weakly ionized plasma of the ionosphere. Somewhere on the front-side of the magnetopause a merging of magnetic field lines from the earth with those imbedded in the solar wind plasma takes place which establishes a way of momentum transfer between the two regions. In the neighbourhood of the central plane of the tail this process is eventually (and in an impulsive fashion) reversed by reconnection of opposing open field lines.

In addition to the large scale transport processes a host of collective microscopic interaction processes of the plasma have been identified. As a medium for propagation and excitation of various kinds of plasma waves the magnetosphere has received intensive attention and also as a laboratory for active plasma and wave experiments. An exciting recent finding is that the earth is a powerful emitter of radio waves with wavelengths in the km range. Fortunately, these waves are shielded from the earth's surface by the ionosphere.

2.2 <u>The Earth, Moon, Planets and Meteorites</u>. In less than twenty years of space exploration we have learned more about the moon, the meteorites and the planets Mercury, Venus, Mars and Jupiter than in all the preceding centuries of earthbound observation. According to Sagan, the Mariner 9 photographic results correspond roughly to

10,000 times the total previous photographic knowledge of Mars
gathered over the history of mankind. The infra-red and ultra-
violet spectroscopic data and other information obtained by
Mariner 9 represent a similar enhancement. The vast amount of
new photographic information involves not only an advance in
coverage or quantity but also a spectacular advance in resolution
or quality. In addition to photographic observation many other
measurements have been carried out in the close neighbourhood of
these bodies, e.g. particle and magnetic field measurements.
Spectacular new views of Jupiter and Mercury and their magnetospheres have been obtained by the latest space missions. All
these informations are needed to understand the origin and the
evolution of the solar system.

The shape of the earth has been determined with very great
precision. Also the elements and molecules in the atmosphere and
their variation with altitude have been measured. Observatories
on space probes showed that Venus is not like the earth as one has
assumed, but that it is hot: at the surface the temperature reaches
750 degrees Kelvin. The atmosphere consists mainly of carbon
dioxide, and the atmospheric pressure at the surface is 70 times
that on the earth. Venus has no magnetic field. This is also
true for Mars, while Mercury and Jupiter have a magnetic field and
also a magnetosphere. The surfaces of Mars and Mercury are very
much like the surface of the moon, namely covered by craters. The
Martian atmosphere is very thin, its pressure at the surface is
only about two-hundredth of the pressure of the earth's atmosphere
at its surface, and it is composed mainly of carbon dioxide.

The planet Jupiter is most interesting due to its strong
magnetic field and its large magnetosphere, which shows considerable
variation; in particular it is sensitive to relatively minor changes
in the pressure of the onstreaming solar wind. Jupiter's electromagnetic environment is further influenced by five of its thirteen
known satellites. Io has a particularly profound effect.

The most detailed informations by space missions have been
obtained from the moon. Photographs, measurements on the moon
and the analysis of lunar material brought back to the earth have
enhanced our knowledge about the moon and its evolution to a very
large extent.

Finally it should be mentioned that also the smaller bodies of
our solar system have been studied with satellites and space probes,
and detailed information is now available about the spectrum of dust
particles and meteorites.

2.3 The Solar Wind. One of the most important discoveries by means
of satellites and space probes has been the continuous corpuscular
radiation from the sun, now called the solar wind, which was first

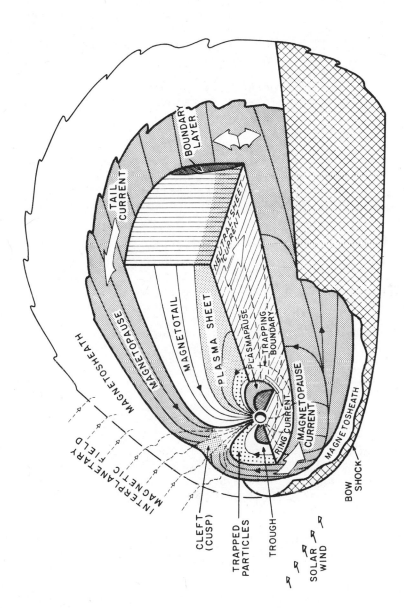

Figure 1. The solar wind, consisting of energetic charged particles ejected from the sun, will deform the earth's magnetic field, sweeping the magnetic field lines away from the sun. A shock front is created, separating the earth's magnetic field from the interplanetary field, and a complicated current system is set up in the plasma inside the shock front. Only very near the earth does the field resemble that of a magnetic dipole. Particles in the solar wind probably penetrate into the earth's atmosphere in the "cleft" regions near the poles. *(Figure courtesy Walter Heikkila, University of Texas at Dallas.)*

suggested due to certain cometary observations and was later also
predicted theoretically.   Near the orbit of the earth and under
"normal" conditions the solar wind has a density of about
5 particles/cm^3, a velocity of about 400 km/sec, an ion temperature
of about 5 x 10^{4}K, and an electron temperature of about 1.5 x 10^{5}K.
The average magnetic field has the form of an Archimedean spiral and
a strength of about 5 x 10^{-5} Gauss.

The solar wind shows considerable variations in density and
velocity and consequently the lines of force of the magnetic field
are also irregular.   Disturbances of the Alfvenic type and propa-
gating shock waves are quite frequent.

The solar wind interacts with the magnetospheres of the planets
and determines their outer boundaries.   The outward flow of the
solar wind will finally encounter the interstellar gas.   The region
occupied by the solar wind in the neighbourhood of the sun is called
the heliosphere.

The solar wind and the imbedded magnetic fields interact also
with the cosmic ray particles coming from the sun and from outside
the solar system.   Their energy spectra and their composition,
especially at energies below 1 GeV/nucleon are strongly affected by
propagation effects in interplanetary space.   Satellite measurements
gave proof of the postulated diffusive nature of the interplanetary
medium, using solar cosmic rays as probes.   The effective diffusion
coefficient through interplanetary magnetic field irregularities is
highly anisotropic*.   The intensity-time profiles and the anisotropy
of many solar cosmic ray events as observed with satellites, revealed
the additional importance of the convective motion of the expanding
solar wind and adiabatic energy loss processes in cosmic ray propagation.

The observed galactic cosmic ray density in interplanetary space
is now known to be determined by a dynamical equilibrium between
diffusive propagation into the heliosphere and convective transport
outward in the solar wind.   In the past the cosmic ray modulation
zone was considered to extend to 10 A.U.   However, recent deep
space probe measurements at a distance of more than 5 A.U. (Pioneer
10 and 11) showed a surprisingly small cosmic ray density gradient
of less than 5% per A.U., which shows that the influence of the sun
on cosmic rays probably extends to more than 30 A.U.

2.4 The Sun.   By space observations a survey of the electromagnetic
spectrum of the sun from the near ultra-violet to the hardest X-rays
has been completed.   The spectrum of the sun is now known over

---

*anisotropic: possessing the power both of left- and right-handed
polarization.

almost the entire frequency range and its variations in time and
with the solar activity cycle are well investigated.   Very briefly,
the results have shown that the quiet sun emits radiation down to
about 100 Å, active regions over sunspots radiate thermal X-rays
down to typically 10 Å.   The spectrum of flares produced in these
active regions is characterized by a grossly enhanced thermal
emission and an extension into the hard X-ray region as a result of
non-thermal electron *bremsstrahlung*.

The solar spectrum has now been explored with sufficient
spectral resolution to identify over a thousand emission lines and
to establish at least relative intensities for many of the stronger
lines.

Images of the sun have been obtained in relatively narrow
bands of energy from 1300 Å down to 8 Å.   These images are important in delineating how the temperature and density of the largest
structures within the atmosphere of the sun vary with depth, position
and time.

Solar studies, from both ground and space experiments, have
revolutionized our concepts of the structure and energy balance of
the solar chromosphere.   The chromosphere outside active regions
is concentrated almost entirely in a network that overlines the
supergranulation boundaries where the photographic magnetic field
is enhanced tenfold.

Some of the most exciting events in astrophysics are those connected
with the acceleration of charged particles and the radiation these
particles emit while interacting with thermal plasmas and magnetic
fields.   In the case of the sun, solar flare studies from space
offer the unique possibility to observe both the fast particles
which escape into interplanetary space and the non-thermal flare
radiation, extending over a wide frequency range, into the gamma ray
region.   The wealth of detail which is observable on the sun is
unparalleled by any other astrophysical object; e.g. the detection
of nuclear gamma ray line emission during large solar flares has
provided direct insight into the energetic processes responsible for
charged particle acceleration.   The signature of these processes is
also imprinted on the chemical and isotopic composition of the
energetic solar particles detected in near-earth interplanetary
space.   Solar studies have revealed the enormous complexity of the
plasma physical processes responsible for fast particle production.
These studies are of interest by themselves but can also contribute
to an understanding of high-energy phenomena of wide astronomical
interest as, e.g. supernova events and explosive processes in
galactic nuclei.

2.5 Cosmic Rays.   Cosmic-ray astrophysics has come into being only
through space observations.   Despite considerable effort, the origin

of cosmic rays is still uncertain, but nevertheless significant information about the nature of the sources has been obtained.

An important step was the accurate chemical-composition analysis of the beam of arriving cosmic-ray particles. The elemental distribution for nuclei between Helium and Iron is strikingly similar to solar system abundances and suggests a thermonuclear origin of the cosmic ray material. The elemental distribution of the very heavy particles beyond the iron group with charges up to Z>90 bears resemblance to the distribution in matter produced in explosive events by rapid-neutron capture. Present evidence is, therefore, frequently taken to support cosmic-ray origin in events of explosive nucleosynthesis. Further clarification can be expected from a measurement of the isotopic distribution of iron-group nuclei which will be available in a few years.

The near absence of the isotope Be^{10}, a radioactive interstellar fragmentation product with $\tau_{1/2} = 1.6 \times 10^6$ years, suggests that cosmic rays reside for more than $\sim 10^7$ years in the galaxy and spend most of the time outside the dense regions of the gaseous galactic disk. Although it is clear that the general interstellar magnetic field is the agent providing storage, the details of cosmic-ray propagation in the galaxy are not yet clear. Here again we can hope to learn the relevant plasma physical aspects from *in situ* observations of energetic particle propagation in the structured interplanetary magnetic field.

The astronomical discipline which relates most directly to the study of nuclear cosmic rays is gamma-ray astronomy. After twenty years of development, the results from two American satellites and most recently from the European COS B satellite, have finally permitted to visualize the galactic disk in the light of gamma rays $\geq$30 MeV. The gamma-ray intensity maps cosmic rays, colliding with interstellar gaseous matter, in a way similar to radio astronomy, mapping the synchrotron radiation of relativistic electrons. Present evidence indicates an increase of the cosmic-ray density towards the galactic centre and a decrease towards the anticentre, a result most easily interpreted in terms of a galactic origin of the bulk of cosmic rays.

Two discrete gamma-ray sources have so far been found, associated with the two fastest pulsars in the Crab nebula and Vela supernova remnant, respectively. These results demonstrate that prolific generation at least of relativistic electrons is currently taking place in these objects.

2.6 The Interstellar Medium. Our basic conceptions about interstellar matter and fields are based on observations from the ground. Major breakthroughs, however, are now expected only from space observations, extending the accessible frequency range into the ultra-violet and infra-red parts of the electromagnetic spectrum.

Of particular importance are studies of dense cloud complexes which are also the places of star formation. The UV absorption-line measurements on board OAO-Copernicus have provided the first direct detection of molecular hydrogen in nearby dark clouds, have yielded the abundances of many inorganic molecules, including CO and the cosmologically important HD molecule. The deuterium/hydrogen ratio in the interstellar gas has also been determined directly from the absorption lines due to the Lyman series of these species, implying a present density in the universe of only 1.5×10^{-31} g/cm^3 if the deuterium is a relic of the big bang element synthesis.

Absorption lines have been detected from low ionization ions of the more frequent elements up to iron and revealed that many of the heavier elements are strongly depleted in dense molecular clouds relative to hydrogen, presumably due to condensation onto dust grains. Cloud properties may also be studied to large distances by means of infra-red continuum and spectral-line measurements. In particular, molecular hydrogen will be detectable by its strong rotational transitions at 17 and 28 µm. A whole new field has developed, dealing with the thermodynamics and chemistry of inter-stellar gas and dust clouds.

2.7 Stellar Evolution. The impact of space science on the astro-nomical disciplines is most spectacular in the field of stellar evolution. While the nuclear-burning stage as the normal life of a star is understood in its principles for about 20 years, space astronomy has provided further insight into the process of star formation and has led to new concepts of the post-nuclear-burning stages of stellar evolution.

Collapsing clouds and protostellar objects have been observed in dense molecular-cloud complexes where dust causes any visible emission to be re-radiated at far-infra-red wavelengths.

The importance of post-nuclear evolution stages has been fully recognized only with the advent of X-ray astronomy. Normal life of a sufficiently massive star ends in a cataclysmic event which may lead to the formation of a compact object, either a neutron star or a black hole. The existence of neutron stars in the form of pulsars is well established. The Crab nebula has been detected as an intense X-ray source and both the Crab and the Vela pulsars as gamma-ray sources. The discovery of compact objects in close binary star systems was one of the major achievements of X-ray astronomy and the case for this object being a neutron star in Her X-1, Cen X-3 and Vela X-1, and a black hole in Cyg X-1 is quite compelling. These objects are powered by rotational or gravitational energy and may reach X-ray luminosities in excess of

the bolometric luminosities of the brightest normal stars.   Thus
the collapse is not the death of a star but rather the transmutation
to a compact state which again experiences a sequence of evolutionary
stages.   Aside from studies in stellar evolution, the X-ray observations of compact objects contribute to studies on the properties of
superdense matter, general relativity, and the plasmaphysical aspects
of the environment of these objects, particularly of neutron star
magnetospheres.   In the case of a black hole, the possibility
exists that the time structure of the X-ray intensity carries the
signature of the metric, mass and angular momentum of the
singularity.

2.8 Extragalactic Astronomy.   As in studies of galactic objects,
extragalactic astronomy has greatly benefitted from the extension
of observations into the infra-red and X-ray region.

Perhaps the potentially most important discovery was the
finding of extended X-ray sources in clusters of galaxies.   The
emission appears to be thermal and indicates the existence of a hot
intergalactic plasma within the cluster.   Its density falls far
short, however, to explain the "missing mass" in clusters of
galaxies and does not provide support for concepts of a closed
universe.

Active galaxies, like Seyfert galaxies, the nuclei of radio
galaxies, and quasars have been detected as powerful X-ray sources.
Some of these objects radiate even more intensely at infra-red
wavelengths, both by thermal and non-thermal processes.   Combined
with optical and radio observations, we have established now a first
experimental basis for an understanding of these extremely energetic
phenomena, possibly involving strange objects as massive black holes
or spinning supermassive stars.

Besides discrete sources, an isotropic background radiation
has been found at radio, microwave, X-ray and gamma-ray frequencies.
While the radio background can easily be explained in terms of
unresolved sources, this interpretation would require significant
evolution in the case of the X- and gamma-ray background with more
numerous and/or more intense sources at early epochs.

A primeval origin is indicated for the 2.7 - 3 K microwave
background.   This interpretation is almost certain now, since the
blackbody character of the spectrum has been established by balloon-
borne microwave spectrophotometry.   These observations have finally
vindicated Gamov's conception of the origin of this radiation and
have strengthened the case for a hot big bang cosmology.   The
remaining uncertainties in the spectral shape are due to severe
atmospheric contamination of the observed signal.   Clearly,
observations from space could reveal the spectral shape to a much
greater detail and detect small derivations from the true blackbody

form, as might be expected from anisotropies or inhomogeneities in the primeval fireball, a prospect which is of great intellectual incitement.

2.9 <u>Fundamental Physics</u>.   Finally, space science has profoundly influenced the development of certain fields in fundamental physics, particularly plasma physics and studies of very dense matter.

The use of satellites has permitted to study the astrophysical plasmas in the earth's magnetosphere and in interplanetary space whose properties largely differ from laboratory plasmas.   The discovery of the radiation belts of the earth and other planets demonstrates that charged particles can be confined by magnetic fields of astrophysical bodies.   In the earth's magnetosphere, particle injection into a magnetic field and subsequent acceleration, storage and loss processes have been studied in great detail along with observations of plasma waves and plasma turbulence.   There is also good evidence that merging of magnetic field lines takes place, a process of wide astrophysical interest.   In the interplanetary medium, the properties of collisionless shock waves, the heating of plasmas by waves and various other aspects of wave-particle inter- actions have been studied observationally.   These results have posed many new problems and have spurred activity also on the theoretical side.   Because of this increased understanding, the near-earth observations are clearly relevant also to studies of astrophysical plasmas not accessible to direct observation.

The study of physical objects whose density is similar to nuclear matter has come into being through the discovery of neutron stars as pulsars and as accreting objects in X-ray binary star systems.   Current problems relate to the composition of matter at very high densities, its physical properties, and the appropriate equations of state;   the physics of neutron star magnetospheres, radiation mechanisms, and the physical conditions in the environment of accreting compact objects.   The explanation of these phenomena will ultimately depend on the work of both astronomers and physicists of many disciplines.

In closing, it should be remarked that space methods could be of decisive importance also to other fields of fundamental physics. The use of satellites and planetary landers could, e.g. provide sensitive tests of the theory of general relativity.

3.   NEW CONCEPTS

For the discussion of the impact of space science, the different findings and discoveries listed above might be grouped into certain categories:

3.1   The earth with its environment could be observed as a planet
like the others; this demonstrated its smallness in the universe
for everybody.

3.2   The observations within our planetary system might be a most
important step for our understanding of the origin of our planetary
system.

3.3   The detection of complicated molecules in the interstellar
space might demonstrate that biological life on other places in
the universe is not improbable.

3.4   Only by the addition of extraterrestrial observations we shall
be able to understand the formation and development of stars.

3.5   This might be true also for our understanding of the origin of
the universe.

3.6   For the observation of certain states of matter (e.g. plasmas,
dense material), extraterrestrial observations are essential.

THE IMPACT OF SPACE SCIENCE ON MANKIND

T. R. Larsen, Norway

An edited summary of the discussion

1.   INTRODUCTION

One of the main driving forces behind space exploration
probably lies in man's urge to know and understand the world in
which he lives.   How did the solar system form?   How did the
universe come into existence?   Is terrestrial life unique or is
it just one example of a universal process?   In the words of the
Norwegian explorer, Fridtjof Nansen: "The history of the human
race is a continuous struggle from darkness to light.   It is
therefore of no purpose to discuss the use of knowledge - man
wants to know and when he ceases to do so, he is no longer man".

But man did want to know.

When launching capabilities for satellites, space probes and
later manned missions and space laboratories became available, man
wanted to use these tools to explore the space, to do research not
previously possible and to make use of space in his daily life
through application of rapidly evolving advanced technology.
Satellites have given us a valuable platform from which observations
can be made and experiments carried out.   We are allowed to look
down towards the earth, do *in situ* measurements, but also concentrate
attention outwards towards the more distant parts of the universe.
Notwithstanding the apparent aspects of military interests, national
prestige, etc., space research has become a good example of a
synthesis between research, application and industrial activity
that has grown strong in many countries.

1.1 Is "Space Science" a Science?   Is there a science which in its
own right can be called space science?   And is it a proper and

13

useful term?   Is it not only a collective for activities using new
tools, new techniques applied to established fields?

At present space science is not a science in the classical use
of the word, but represents interdisciplinary scientific activity
of great scope.   Plasma physicists, astronomers, earth scientists,
meteorologists, oceanographers and scientists from a number of other
fields work together using an advanced technology in rocket,
satellite and space probe research.   Space science meets
intellectual challenges in very dynamic areas where boundaries
between the subjects change or become meaningless.   Radioastronomy,
earlier regarded as an independent field, is now quite closely
integrated with astronomy and astrophysics.

Space science may in the future come to mean the new astrophysics.   This field has now left the speculative stage and become
an experimental undertaking to a much larger extent than previously.
The *in situ* measurements in interplanetary space have given the
first experimental understanding of the properties of plasma in
space.   The study of magnetospheric phenomena will be the basis
for plasma studies in the whole universe.   This new information
and the possibility to make observations throughout the entire
electromagnetic spectrum indicate that one here is faced not so much
with the continuation of the present astrophysics as with the birth
of a new science.

One major obstacle, however, for this view to become accepted
is a "mental blockage" due to the presence of ideas and thoughts
developed earlier.   These ideas with their tremendous inertia may
be preventing us from interpreting the new observations in the
correct way.

1.2 What Is Understood by "Impact"?   In the context of the Symposium
"impact" means "a compelling effect which changes our way of thinking
or leads to changes in our actions".   This process, however, is an
evolutionary one and one should be aware of both the impacts that
have already become apparent as well as those that manifest themselves only in the future.

One should take note of the distinction between impulse and
impact.   In the context of scientific and technological development
an impact can be said to have occurred when there is not only a
change in "acceleration" in a certain field, but also a change,
often unexpected, in the direction of its evolution.   In our framework an impulse would not cause the evolution to take new directions.

An "impact" should not be measured by its temporary
"popularity".   Such an interpretation can only confuse the issue
under discussion; we are here concerned with the profound impacts
and their persistent consequences.

1.3 Through Science to Mankind.   Since the real impact of new
discoveries may become obvious only over a long-term perspective,
one should give the word "Mankind" a liberal definition.   In our
discussions the word generally does not mean the public as a whole,
nor the majority of the people, nor even a majority within the
scientific community, but it may mean only a small group of
scientists with deep knowledge of the specific problems concerned.
Inherently, scientific results are initially of most importance to
the scientists who are active in the particular field.   History
has shown, however, that the impact of scientific discoveries in
due time has often had fundamental influence upon the lives of all
men.   The example of Copernicus can be mentioned.   His thoughts
probably had impact only upon a small group of people in his time;
yet his thoughts were revolutionary in a historical perspective.

2. IMPACTS OF SPACE SCIENCE

Visionary powers would be required if we today were to select
the impacts which in the course of history will prove to be of
genuine importance to mankind.   This summary strives in an implicit
way, however, to convey the importance that the learned participants
at the Symposium attached to the various scientific results of space
science in light of our current knowledge.

Many of the burning questions which were in the minds of scientists twenty to thirty years ago have been solved or progress towards
their solution can be detected.   Man has learned more, but as a
result, he is humbly aware also of what he does not know.   What
then are some of the most outstanding achievements in space science?

It is here important to keep in mind that the following
discussion is concerned essentially only with space science, which
is but a small part of the activities one usually understands by
the "space programme" (e.g. in the United States less than 25% of
the funds for the space programme is used for scientific purposes).

2.1 The Scientific Impacts.   The first two fundamental scientific
impacts are inherently related to the onset or birth of space
activity.   Today, it sounds almost trivial to mention that it now
is possible with a rocket, satellite or space probe to bring
recording instruments to the places in space where we want the
measurements to be taken.

Prior to the historic launching in 1957 of Sputnik 1, the first
artificial earth satellite, geophysical research of the earth and
its environment, however, had been limited to the use of ground-
based techniques only supplemented by balloon and sounding rocket
observations which became feasible after the end of World War II.

Furthermore, up to the 1950's and early 60's all knowledge that man meticulously had gathered experimentally about the universe had been based on the study of radiation in the electromagnetic spectral regions that penetrate the earth's atmosphere.   As is well known in only two spectral ranges does the radiation reach the earth's surface without severe attenuation, namely one in the visible range extending into infra-red and one at radio wavelengths.   These restrictions have imposed fundamental limitations upon all efforts of acquiring knowledge about our solar system, our galaxy and regions more distant in the universe.

The conquest of space made it possible to circumvent these two basic limitations, but the full impacts of these possibilities are not yet explored.

3.   IMPORTANT   NEW   DISCOVERIES

In retrospect the brief history of space science exhibits a number of important single discoveries or ensembles of findings which stand out as particularly important.   Any listing of a few of these is open for debate;   one selection presented at the Symposium contained the following three groups of achievements:

1)   The detection of the earth's radiation belt and the subsequent findings from the investigations of the earth's magnetosphere.

2)   The detection of the solar wind and the investigations of the properties of planets and of the interplanetary space.

3)   The discoveries made in X-ray astronomy.

In discussing these points it was maintained that magnetospheric research had brought forth a much better understanding of the earth's environment.   The understanding of the physical processes taking place will have profound influence upon our picture of the solar system as well as the universe as a whole.

The detection of the solar wind was important since it is probably a phenomenon associated with all stars.   The interplanetary plasma is the only cosmical plasma directly accessible to man and will provide information to allow for much more reasonable extrapolations to galactic plasmas than would be possible only from laboratory experiments.   Concerning the atmospheres of the nearest planets in many cases direct *in situ* measurements have shown that numbers derived from earth-based measurements of planetary entities were completely wrong.   The solution of general problems of the circulation and energy balance of planetary atmospheres were pointed out as having far-reaching importance.   As a result of the study of the geologies of the moon, and the nearest planets, one will be able

to approach the origin of the solar system in a new way.   In a not
too distant future it may thus be possible to give a better descrip-
tion of the history of planetary system.

The X-ray astronomy may prove to become an entirely new field
of astronomy almost comparable to radioastronomy.

3.1 If the view is shifted towards the long term scientific
impacts, other future possibilities were emphasized, particularly
in relation to the information they could convey about the universe.
The available "information channels" may be listed as follows:

    the electromagnetic radiation
    cosmic radiation
    neutrinos
    gravitational quanta.

Particularly the first two could give most important information.
The study of cosmic particles covering eighteen orders of magnitudes
in energy is still in its infancy, but is a field which should be
given very high priority.

4. THE LAWS OF PHYSICS AND COSMOLOGY

From a scientific point of view the space activities have
contributed to the fundamental quest: are the laws of physics
invariant over the universe or are there small or large scale
variations with space.   As far as we know, the laws of physics
are invariant, i.e. throughout the universe they are the same as
determined on our local earth.

Another related problem regards the invariance of these laws
with time.   It is important to establish whether time scales other
than linear may be more meaningful.   The laws of nature are
invariant with time as we know them today; however, we may
discover phenomena that cannot be explained on a linear time scale.

New intriguing fields of physics have been discovered by radio-
astronomic means or by satellites; only keywords were cited;
pulsars, black holes or antimatter.   The field of condensed matter
may contain important information for our understanding of the
universe and may offer a test of the validity of the general theory
of relativity.

The isotropic 2.7K black-body radiation is but one of the new
findings that must be explained by any cosmological theory.

5.   IMPACT UPON OTHER SCIENCES

In using the space as a large natural laboratory one has been able to study atomic and molecular processes which cannot at present be studied in earth-based laboratories.   Plasma physics is possibly the field which has benefitted the most from space science, but the impacts upon atomic, molecular and solid state physics have been significant.

The advent of space science allowed us to view the earth from outside and has given us new knowledge.   Studies of the orbits of the early satellites gave information about the earth's gravitational field and about the shape of the earth.   The knowledge about the earth's inner core and the source of the geomagnetic field is still inadequate.   In light of the importance of planetary magnetospheres for the preservation of the atmosphere this subject has received new significance.   It is conjectured that during periods of geomagnetic reversals, the earth's ozone layer is partially depleted through an increased production of NO due to the more easy access of energetic particles from the sun into the earth's atmosphere.   Would life on the earth be affected?

A number of other important impacts of space science were discussed later during the Symposium in connection with the treatment of space applications.

6.   OTHER IMPACTS

An impact which does not come so much from the space science proper as from the space programme, is the experience gained by managing a vast programme.   This knowledge may be of genuine importance in the organizing of the future large projects, e.g. in a world-wide undertaking against hunger.

Various industries have gained significantly from space science through technological developments.   A thorough elucidation of the technological impacts were not presented at the Symposium, although several speakers in general comments referred to e.g. the development of the electronic computer, the semiconductor technology, the new material technology and more exotic techniques like crystal growing in space and zero-g technology.   The importance of electronic technology in medicine was also pointed out.   The field of quality control has been developed to such a degree that reliabilities, which one ten to fifteen years ago could only dream of, are today obtainable.

7. LIFE ELSEWHERE IN THE UNIVERSE?

An important effort is the quest for life elsewhere in the universe. The looking for origins is an exciting, but extremely difficult task. Many of us were perhaps surprised to learn about the complex organic molecules in interstellar space. When the chemistry and physics related to such molecules were done properly, it turned out, however, that the existence of such molecules was not improbable at all.

Can space science in its search for life shed some light upon such fundamental properties as cell growth and cell multiplication?

There is today a growing feeling of near certainty statistically speaking, that life exists at numerous other places in the universe. It may be a common property of star formation to also develop planetary systems. And the conditions present on the earth some billion years ago when life evolved, are probably not unique in the universe.

What will it mean for man's image of himself if he learns that life exists elsewhere in the vast space surrounding his "space ship", the earth? Will it be too much to bear or will mankind accept such knowledge as a natural manifestation of a cosmic property? Exobiology will certainly enter into a new era if traces of organic life are detected on Mars by the Viking missions in 1976.

8. IMPOSITION?

Are expensive tools being forced upon the scientists – tools they have not asked for and which they do not want? Or do the scientists lack perspective to see the possibilities opened up by the new technology?

It is probably relevant to make the point that the space programme was not initiated for scientific purposes; the initial interests were of military nature, subsequently prestige came in as another driving mechanism.

Admittedly, the tools were therefore available. In the early days of space science, there was great enthusiasm and data of significant scientific value were obtained using German V-2 rockets. It is fair to say that the scientists made use of the possibilities that were opened up in a traditional, but sound way.

Today, when the elaborate "Spacelab" is becoming available, many scientists seem to be hesitant. Is this reaction due to lack of perspective, due to inhibitions from a bureaucratic machinery,

or due to a feeling that the most important experiments have already
been made?   A plea was made for the bright ideas and a step increase
in mental effort.   Space science needs a genius;   the Spacelab does
not at the moment bear the stamp of original thinking.

Several speakers argued that Spacelab was imposed upon the
scientists who were asked to provide good ideas and build equipment,
whereas many scientists wanted to proceed in other directions, do
other types of experiments.   Even though such experiments would be
cheaper than Spacelab experiments, adequate funding did not seem to
be available.

The complexity of organizing a major scientific space project
was also pointed out.   It takes often ten years with planning,
testing, experimenting and data analysis before the results can be
presented.   In such a programme a failure can be disastrous for
the career of the scientist involved.   In "big science" another
important task is to insulate the inventive scientist from becoming
too involved with management.

Today's space science would therefore very likely not attract
scientists of Rutherford's type.   The pioneering days are over and
during the natural process of maturing, space science does not offer
the same intellectual excitements as earlier.

9.   TIME   WAS   RIPE

The question was raised whether mankind needs some kind of
target larger than the day-to-day problems.   Earlier challenges
came from wars.   Were such thoughts in President Kennedy's mind
when the goals for the moon programme were set?   And if society
for some reason, be it economic or of another type, needs a
national goal to strive for, was the space programme the best way
to achieve this?   Can it be said that the moon programme was meant
to be the "pyramid" of the twentieth century?

A legitimate question to ask is whether a vast programme for
cancer research would have been a more worthwhile project?   Or a
project to eliminate hunger in the world?   There seemed to be a
wide concensus at the meeting that such programmes are really not
alternatives;   x million dollars are not necessarily forthcoming at
the same time for any such programme.   The underlying reasons
behind the attraction of one project over another were not spelled
out, but several speakers acknowledged that time was ripe for the
space programme in the early 60's.   A positive political climate
and an adequate scientific basis which give hope for success, are
but two of many necessary prerequisites for a large programme to
get started.

10. LACK OF SCIENTIFIC IMPACT

In space science experimentation involve data-gathering measurements often continuously over long periods of time and of a large number of objects.   The philosophy of such experiments is different from that connected with other sciences;   in the nuclear research at CERN to take one example, the scientific goals may be better defined and less data are needed.

There is obviously a necessity to be more discriminative in space research.   All the data which are collected may not be needed;   new data-handling techniques, however, may be needed to remedy the situation.   One must find a balance between obtaining too much data on the one hand and precluding new discoveries due to too much pre-programming on the other.

The developments so far have not taught us how to cope effectively with large data masses;   when exploration in connection with earth resources are started more systematically a data "explosion" may be in sight.

There was furthermore a general dissatisfaction with the present lack of impact of space science upon education.   As a basic scientific endeavour space science can widen our knowledge of human life and its surroundings and thus be a part of our quality of life. Criticism was voiced in relation to the large time lag between the updating of textbooks.   The scientific results are not disseminated in the proper way.   Popular accounts of the achievements too often lack the original inspiration with which the new knowledge has been uncovered.

11. THE IMPACTS UPON THE PUBLIC IN GENERAL

Several speakers argued that it was still far too premature to look for an extensive impact of space science upon mankind, especially if one interpreted mankind to mean the general public. It was recognized that there had been major scientific impacts, but one had to await the "practical" impacts before one could discern any impacts upon the average citizen.

In the discussions this view was not challenged, but a few examples of events which did receive world-wide attention and possibly created a deeper appreciation were offered.   The pictures of the earth taken from space were mentioned in this connection. For the first time could man see the earth as one of the planets and the pictures provided a pertinent reminder of the earth's limited size.

12.   IMPACT   UPON   POLITICIANS

The space programme could not have been carried out without
the ideological and financial support of politicians in many
countries.   The quality of political leadership in relation to
space science and science in general was recognized.   The subject
of financial support had been lifted above a popular gallup and the
results of space science have been favourably received among most
politicians.

It is the fault of the scientists themselves if one does not
find in society in general an understanding, realization and
appreciation of the results of space science.   We as scientists
may have erred;   in order to be in harmony with the society one
cannot depart too far in our role as specialists.

13.   A RAY OF HOPE OR A CRY OF DESPAIR?

To date the space activities have not had a direct impact on
the environmental movement nor did the space programme give rise to
it.   This statement was not contested although it was claimed that
the Apollo programme acted as a catalyst on dormant knowledge and
concern for the environment.   The moon landing may have prompted
some people to realize that there are many unsolved problems
remaining on the earth.   It may also be that this enormous
technical feat has given nourishment to the hope that mankind may
now be able to solve these problems.   There may thus be a message
of hope in the words: "If you can put a man on the moon, why cannot
you......"?   The "impossible" is no longer impossible.

This interpretation was challenged.   The agnasciam does not
represent a ray of hope, but is a cry of despair.   It is a reaction
over the neglect of every-day problems and over the large spending
of money involved.   Viewed from this angle the "cry" is a beg for
greater influence upon the setting of priorities for capital
projects of governments.   The impressive moon-landing thus gave
truth to the saying that "travelling safely is better than
arriving".   The spectaculars gave rise to "anti-impacts".

The question pertaining to what degree science and technology
is beneficial for society was, however, left unanswered along with
many other questions.   The Symposium did not provide an extensive
list of merits for the first two decades of space science.
Attention was rather focused on the more outstanding areas from
which impacts were supposed to originate.   The Symposium gave a
reflected evaluation of the space science endeavours with no
inclination towards "overselling" of the achievements.

This intellectual objectivity could not hide the fascinating
and exciting perspectives that lie ahead.   Entirely new facilities
for experimental studies will become available.   Today, we are
probably only at the onset of an evolution which may be more
important for mankind than the revolution that came with Copernicus.
Just as important as the interdisciplinary character of space
science is in bringing scientists from many fields together, is the
international aspect in meeting people from many nations in a common
search for new knowledge and deepened understanding.   In the pursuit
of the secrets of our universe national differences and boundaries
become meaningless and unimportant.

May we here really detect a ray of hope?

SATELLITE COMMUNICATIONS

J. V. Charyk

Communications Satellite Corporation, Washington, D. C.

1. INTRODUCTION

The last decade has seen the addition of a new dimension to
our global telecommunications capability, viz. the communications
satellite. In a single decade we have reached the stage where the
majority of all long distance international traffic is carried by
satellite. Domestic and regional systems have entered into
service and we are on the threshold of using satellites for communications with mobile platforms on a regular commercial basis. Ahead
lie new and exciting possibilities which give strong promise of
exerting a major impact on the nature of the world of tomorrow. It
is therefore most appropriate that under these auspices we review
the progress to date. We can then, with the advantage of this base,
seek to focus our thoughts on the problems raised by these technological developments but perhaps more fruitfully on the potential of
their developments in the world of tomorrow. Although many of you
may be familiar with the developments in the communication satellite
field in the past decade, it appeared useful to put together a com-
pendium which demonstrates its rather dynamic character and provides
a common base for our discussions here in trends, impact and
potential.

Other applications such as the maritime, aeronautical, and
broadcast satellite service are in about the same stage as international fixed service was ten years ago, but with that experience
behind us the technical success of these ventures seems assured,
although the political and economic aspects are not as clearly
foreseen.

2.   INTERNATIONAL   FIXED   SATELLITE   SERVICE

2.1 <u>Organizational Background</u>.   Since communication is, by its
very nature, at least a two-party operation, an international
communications system is only useful if the communications entities
of other countries agree to use it and to provide appropriate
terminal and distribution facilities to handle the traffic.
Accordingly, shortly after COMSAT's formation in February 1963,
visits were made to a number of countries around the world.   There
was great interest in this new venture, but it was tempered with a
certain amount of incredulity and concern at the very rapid development and implementation programme proposed.   Through numerous
meetings and informal discussions, there emerged over a period of
months a confidence that an economically viable communications
satellite system at an early date was, indeed, a possibility, and
a desire to join as full partners in this unique venture.   This
was formalized in August 1964 in a pair of agreements - the first
signed by the governments of all the participating nations, and the
second signed by the designated telecommunications entities of these
countries.   For the United States, this was the Communications
Satellite Corporation.

The first document spelled out the desire of the signatories
to establish a single, global commercial communications satellite
system as a part of an improved communications network which would
provide expanded telecommunications services to all areas of the
world and which would contribute thereby to world peace and understanding.   It enunciated the belief that satellite communications
should be organized in such a way as to permit all states to have
access to the global system, and to those states so wishing, to
invest in the system with consequent participation in the design
and development, construction, establishment, maintenance, operation
and ownership of the system.   It established the goal of producing
such a single global commercial communications satellite system at
the earliest practical date.   Nine governments signed the documents
in August 1964.   That number has now increased to ninety-one.   By
April of 1965, the first satellite was successfully launched.

The agreement set up a committee as the top policy body of the
international joint venture.   This committee was called the Interim
Communications Satellite Committee, and the parent organization the
International Telecommunications Satellite Consortium, or INTELSAT.
A country wishing to join would pay an amount related to its share
of international telephone traffic, and its vote would be based on
its investment.   Each entity using the satellite system, whether a
member or not, would pay a specified amount per channel per year for
use of the space segment.   The amount was set so that it would
cover anticipated satellite costs and return a profit to its members.
While the space segment was jointly owned by the INTELSAT members,
each was responsible for its own earth stations including its

purchase, or construction, and operation and maintenance.

COMSAT was designated by the signatories as Manager for the design, development, construction, establishment, operation and maintenance of the space segment.   Thus, with a charter established and a business objective spelled out, a major step was taken towards establishing a global communications satellite service.

The interim arrangement outlined above was superseded on February 12, 1973 by a definitive arrangement, the product of extensive negotiations over a period of about two years.   The original two-tier organizational structure, of a Committee and a Manager, was replaced by a four-tier structure consisting of an Assembly of Parties (member governments);   the Meeting of Signatories (the operating telecommunications entities);   a Board of Governors;   and an Executive Organ reporting to the Board of Governors.   The Assembly of Parties meets every two or three years and considers general policy and long term objectives.   The Meeting of Signatories is an annual convention and considers the annual report and financial statement submitted by the Board of Governors and sets the minimum investment share for a seat on the Board of Governors.   The board is made up of about two dozen governors who represent countries, or groupings of countries with at present more than about 1.1% of the shares, and representatives of geographical regions.   They meet about six times per year and are responsible for all facets of INTELSAT's space segment.   The Executive Organ is headed by a Secretary General responsible for management services of an administrative and financial nature.   By December 31, 1976 he will be succeeded by a Director General who will be responsible to the Board of Governors for all management services.   With the entry into force of the definitive agreement, COMSAT was given a management services contract for the performance of INTELSAT's technical and operational management services, terminating February 11, 1979.   By the end of 1974 INTELSAT members had invested $308 million (net) in its global satellite system and had seen it become the major means of international communications.

The Intelsat concept was as radical an approach to international organizations as the satellite to international communications.   Its success may well see it used as an example for other international undertakings.

In the era of H.F. transmission, large fixed directional antenna arrays were used, each aimed at one of the various regions with which it was desired to communicate.   Since the amount of communication was relatively small, largely because of the poor quality of transmission, it was uneconomic for all countries to put up separate arrays for all possible regions, and therefore when necessary it was customary to relay traffic via a third country which had antennas

aimed at both regions.   In Africa and Asia, where the antennas were
often installed by a European country, they were aimed only at that
country forcing all communications to other countries to transit,
with a charge for both halves of the circuit.   The satellite
changed all that.   An earth station can be purchased and installed
at a cost of $3 to $5 million, and high quality telephony and TV
circuits to all parts of the world immediately become available.

2.2   Technical Background.   The major technical decision to be made
in 1963 was whether the first satellite system should be based on
the "medium altitude, random orbit" approach or on the "synchronous
orbit" approach.   In the first case, some two dozen satellites
would be randomly disposed along an orbit ten to thirteen-thousand
kms above the earth, while in the latter case a single satellite in
an equatorial orbit thirty-six-thousand kms above the earth's
surface would be used, so that to an observer on the earth it would
appear to be fixed in space.   The medium altitude system had the
advantages of a simpler satellite design not requiring position
control jets and fuel;   less transmission delay of the signal;   a
single system of satellites could be used simultaneously throughout
the world;   and graceful degradation in that a failure of one
satellite of the two dozen had a negligible effect.   Its disadvan-
tages were that two dozen satellites were needed against six for
the synchronous system (one per ocean area plus a spare for each)
and that two antennas were needed per earth station to avoid
service interruptions.   One antenna would be required to pick up a
satellite coming into view while the other was terminating service
on a satellite going out of view.   Even so, in a random system
there would be regular outages, albeit on a predictable basis.

     The synchronous satellite approach raised the question of the
effect on the telephone subscriber of the approximate quarter
second delay for the signal to be transmitted from one station to
the other.   At that time it was unclear whether a delay of this
amount would be acceptable to the user.

     An important point that emerged from many discussions on this
choice of orbit was that a successful launch of one or two medium
altitude satellites, which would be required to prove out the design
before committing to a launch of two dozen satellites, would at best
still be an experiment since the two satellites would only be in
view for short periods a few times a day.   The successful launch
of a synchronous satellite, on the other hand, would permit full
operational use after an experimental phase to determine the problems involved in keeping the satellite positioned in space, and
the acceptability of the transmission delay.   It was therefore
decided to concentrate on developing and launching a synchronous
"experimental/operational" satellite (Early Bird).

With the successful launch of Early Bird in 1965, extended tests were made of the effect of the time delay on the subscribers. Such tests, run over a period of many months by more than a half dozen countries and communications organizations, verified the suitability of a synchronous satellite system to provide quality international telephone services.

2.3 <u>Present Status</u>.   Sixty-five of INTELSAT's ninety-one member countries now operate ninety-one stations including one-hundred-and-fifteen antennas.   While satellites are jointly owned by the member countries in shares proportional to their share of international traffic, the earth stations are wholly owned by the administrations in the countries where they are located.   The amount of traffic per station varies widely from a handful of circuits to over fourteen-hundred circuits per station; from a single path to about three dozen paths per station.   As of mid-1975, these stations were carrying a total of over six-thousand-five-hundred two-way telephone circuits, and over three-thousand-five-hundred hours of TV programming per year* via the more than three-hundred-and-eighty earth station-to-station paths** now in service throughout the world.

Of the ninety-one stations in use, the lowest half generate only ten per cent of the traffic, while the highest ten per cent of the stations generate fifty per cent of the traffic.

2.4 <u>Satellite Development</u>.   Four generations of satellites have been used by INTELSAT during its first decade from 1965 to 1975. The smallest satellite, Intelsat I, weighed eighty-five pounds, and could provide two-hundred-and-forty two-way telephone circuits but only between two countries both of which had to be in the northern hemisphere.   It had sufficient power to utilize about ten per cent of the available bandwidth.   Intelsat II occupied twenty-five per cent of the available bandwidth and also provided two-hundred-and-forty telephone circuits, but these could be distributed among a multiplicity of stations, both in the northern and southern hemispheres.   Intelsat III, weighing some three-hundred pounds was the first to use a directional antenna which illuminated all of the globe visible from a synchronous satellite, about one-third of the total.   The increased power available on this satellite plus the directional antenna permitted the satellite to utilize the full bandwidth of 500 MHz and provide fifteen-hundred two-way telephone circuits.   Finally, Intelsat IV, the present work horse of the INTELSAT system, provides some four-thousand two-way telephone circuits and weighs sixteen-hundred pounds.   It also uses the

---

*   TV transmission accounts for only about 2% of INTELSAT's gross, about the same as for cable restoral, while 95% is due to telephony, telegraphy, and data transmission.

** The number of country-to-country paths totalled 523.

full 500 MHz bandwidth but obtains the increased number of channels by using more transmitter power than in Intelsat III and also by using directional antennas which cover one-fifteenth of the visible globe, thus providing still higher power in those areas.   All of these satellites have been spin stabilized.   More recently, satellites using three-axis stabilization have been successfully flown and operated at synchronous altitudes and provide certain advantages to the designer.   These include more precise pointing accuracy and also greater flexibility in trading off weight and power in the satellite design.   It is expected that future satellites may well tend to go in this direction.

2.5   Methods of Increasing Channel Capacity.   In the first four generations of satellites the constantly increasing demand for more channels was met by increasing the usable bandwidth and/or power in the satellite.   In the Intelsat IV this trend reached its practical limit;   for example, connecting a satellite transponder from a global beam to a spot beam provides an increase of radiated power of approximately fifteen times, and would provide a fifteen-fold increase in channel capacity if a fifteen-fold increase in bandwidth were available, but since the satellite already uses all the available bandwidth from 3700-4200 MHz a further increase in this direction is not possible, and the fifteen times increased power only doubles the capacity.   If another doubling of channel capacity were desired solely by increasing power, it would have to be increased about an additional seventy-two times rather than fifteen times.   Clearly, this approach is very inefficient and therefore much effort has been expended on alternate means of increasing channel capacity.

The first approach to obtaining greater capacity for a given spectrum allocation has been used in the Intelsat IV-A (see Fig. 1) the first of which was successfully launched in September, 1975. In this satellite, eight transponders are connected to an antenna beam which illuminates Europe and Africa while another eight transponders connect to another antenna beam which illuminates South America and the eastern part of North America, using the same frequency band as that used for the first eight transponders.   The two beams are sufficiently separated so that the amount of interference between them is reduced to a negligible value.   An additional four transponders* are connected to a global beam which illuminates all of the above continents and is used for transmissions such as TV which are usually aimed at all of these areas;   and also for communications to the islands in the Atlantic which are not covered by the narrower beams.   By this technique the capacity becomes approximately six-thousand-five-hundred two-way telephone

---

* Using one-third of the allocated 500 MHz;   the other two-thirds being used (twice) by the two sets of eight transponders connected to the continental beams.

Figure 1 - INTELSAT IV-A Satellite

circuits as compared to the four-thousand of Intelsat IV.

Another method of re-using the frequency band is through "dual polarization". In this approach the several amplifiers of a satellite all connect to an antenna whose radiating element is arranged vertically to the satellite. A second set of amplifiers using the same frequency band as that of the first set, connects to an antenna which is horizontal. Tests indicate that properly designed antennas both on the satellite and the corresponding antennas of the earth stations, can be designed so as to permit operation with a minimum of interference between the two polarizations. A system of this type is incorporated in four satellites being developed by COMSAT for AT & T's use in their domestic network, the first of which is to be launched in early 1976. Through this technique a doubling of satellite capacity can be achieved. This concept will also be incorporated into the future Intelsat V system.

Still more additional capacity can be obtained by using new frequency bands which were made available to the fixed satellite service in the World Administrative Radio Conference (WARC) in 1971. These bands are in the 11/14 GHz and 20/30 GHz regions. The bandwidth of the former is the same as that of the present 4/6 GHz band; i.e. 500 MHz in each direction from earth to space, and space to earth. The 20/30 GHz band has five times this width and is therefore much to be preferred, from bandwidth considerations alone. Unfortunately, this pair of bands is affected by rainfall and in the case of the higher band it is no longer practical to provide sufficient satellite power to override the effects of heavy rains. However, by utilizing two stations some ten to twenty miles apart, it should be possible to provide continuity of service. At these high frequencies not only does one have the advantage of greater bandwidth but in addition a narrow beam can be produced with a relatively small antenna. For example, 2 m diameter antennas will produce a beam width of 0.5° at 20 GHz. This raises the possibility of designing a satellite with a number of such narrow beams to permit a corresponding amount of frequency re-use. It then becomes necessary to provide means of interconnecting the repeaters for these beams on board the satellite, either by means of a time division switch (for digital communications), or a matrix of frequency filters, and such techniques are being developed.

Still another approach to the problem of obtaining increased channel capacity is by the use of time division multiplex. Such methods have been used in terrestrial systems since the early 60's and they have definite advantages for satellites as well. Since the earth station owner in INTELSAT pays all of the cost of his earth station equipment but only a fraction of the cost of the space segment, the tendency is not to go to any new techniques requiring

new equipment in his station.   The use of digital techniques permits
simple, effective means of utilizing the gaps in the speech when one
party listens to another thus resulting in a doubling of capacity.
INTELSAT is planning field trials of such digital time division
equipment later in this decade.

The SPADE system has been designed for countries with relatively
light traffic to a number of other countries.   It is based on demand
assignment, a single radio frequency carrier for each voice channel.
At the present time some sixteen countries are using this system.
This technique also doubles the capacity of a given transponder.

One obvious means of increasing the system's capacity is to add
additional satellites.   This has been done in the Atlantic region
where two satellites are now in use.   The price paid is that some
of the countries then require two antennas - one for each of the
two satellites.   One means of improving on this problem is a
multiple beam antenna such as the Torus (Fig. 2).   This antenna,
developed by COMSAT Labs, incorporates multiple feeds so that a
single fixed structure can be used for operation of two or even
three satellites simultaneously.

While the first of the Intelsat IV-A series of satellites has
just been launched, five more are under construction and INTELSAT
has requested bids for its follow-on series of Intelsat V satellites.
These are planned to provide twelve-thousand-five-hundred two-way
telephone circuits by incorporating the dual beam concept of the
Intelsat IV-A series;   the use of the new frequency bands 11/14 GHz
in addition to operation at 4/6 GHz, and the use of dual polarization for the 4/6 GHz region.

2.6   Earth Station Considerations.   As with satellites, earth
station design has evolved through a number of generations during
the last decade.   About ten years ago station costs ranged from
$6 to $12 million, but several years later dropped to the range of
$4 to $8 million.   In 1970 the prices were $3 to $6 million and by
1972 the cost of typical stations ran in the range of $2.5 to $5
million.   The inflation of the last two years has increased this
last range to $3 to $6 million.   These stations differ from terrestrial microwave stations mainly in three areas.   The antennas are
30 m in diameter as compared to 2 to 3 m of a terrestrial radio
relay station;   the earth station transmitters range from hundreds
of watts to several kilowatts as compared to several watts;   and a
low noise receiver uses a refrigerator operating at about 15° above
absolute zero as compared to a simple crystal mixer used in terrestrial radio relay stations.

For countries needing but a handful of circuits it becomes more
economic to use an antenna of 10 m diameter and pay the increased

Figure 2 – Torus Antenna

space segment cost.   With standard large earth station antennas
the space segment cost is now $8,460 per year per one-way telephone
channel;   whereas when using a 10 m antenna the space segment cost
increases by a factor of 2.5 since approximately that much more
satellite power is needed;   however, with less than a few dozen
circuits the overall costs can be decreased by using the smaller
antenna.

During 1974 the ninety-nine earth station antennas and related
transmitters, receivers, etc., in the INTELSAT system had an average
continuity of service of 99.95% corresponding to an average total
outage of less than 4.5 hours per year including outages due to
planned maintenance.   During that year twenty per cent of the
systems had outages less than one-tenth of this amount, that is,
99.995% corresponding to an average total outage per station of
less than twenty-six minutes per year.   Such performance compares
favourably to that of present terrestrial communication systems and
considering the complexity of these earth stations and the fact that
a majority of them have been in operation only a few years such
performance is indeed outstanding.

3.   AERONAUTICAL   SATELLITE   SERVICE

3.1 Background.   Studies made as early as 1964 established the
technical, but not the economic, feasibility of using a satellite
as a repeater for communication between aircraft and earth stations.
Confirmation was obtained with successful experiments first with
teletype, and later voice, using NASA's SYNCOM III, ATS-1, and ATS-3
satellites.

A series of ICAO-sponsored meetings were held over the years
and gradually the key issue raised was that of choice of frequency,
whether it should be in the VHF (118-136 MHz) or in the L-Band
(1540-1660 MHz).   This difference of opinion was not a purely
technical problem but involved economic and national issues.   The
U.S. airlines advocated the VHF band with which they had many years
of experience, while the Europeans were the main proponents for the
L-Band system because it was not as crowded as at VHF, and because
it had the potential of 0.1 mile position-fixing accuracy compared
to one mile at VHF.   The airlines argued that there was sufficient
room at VHF, that it did not really matter, for all practical purposes, whether the accuracy was one mile or 0.1 mile, and that the
advantage of using simpler, proven VHF equipment was a most important factor.   An attempt to promote development of a wideband
satellite with equipment for both bands failed when the U.S.
declared for L-Band in January 1971 over the objections of many
airlines.

3.2 Present Status.   After three years of further discussions an

agreement was reached among the European nations, Canada and the
U.S. under which an aeronautical satellite project ("AEROSAT")
would be jointly funded in the ratios of forty-seven per cent, six
per cent and forty-seven per cent respectively. The Europeans
would be represented on a council by ESA (the European Space
Agency); the U.S. would be represented by COMSAT; and the
Department of Communications would represent Canada. The satellites
will include both L-Band and VHF band equipment but the cost of the
latter would be completely charged to the U.S.A.

In early 1975 a joint team began preparing specifications for
development and construction of several satellites. These will go
out for bid by teams of companies from the various participating
nations.

There remain differences of opinion between air traffic control
and airline representatives insofar as the value of satellite services is concerned. The former see importance in a two-satellite/
ocean surveillance system to permit reduction in aircraft separations
while the latter feel the only requirement is for improved communications.

From a spacecraft viewpoint, the proven feasibility of unfurling
large antennas in space, as for example the 9.25 m diameter para-
bolic reflector of ATS-6, opens the way to the possibility of
illuminating the airways with a multiplicity of narrow beams
rather than a large single beam, thus providing the possibility of
frequency re-use and a greater number of channels and/or simpler
radio equipment in the aircraft, or a combination of both.

The availability of sufficient high quality circuits to transoceanic aircraft will spur the introduction of connecting the planes
to the national telephone networks for passenger communications.

There is little doubt but that the next decade will see aeronautical communication via satellite as the widely accepted means
of transoceanic aircraft communications.

4.   MARITIME   SATELLITE   SERVICE

4.1 Background.   The application of communications for ship-shore
use via a satellite repeater was first studied about ten years ago
at COMSAT.   The results indicated that such communications could
be implemented with available technology but the economics of the
project did not justify a commercial venture.   The technical simi-
larities between the communication satellite systems for maritime
and aeronautical use were sufficiently close to warrant investigating the possibility of a joint service.   However, at least thus
far, both groups prefer to keep their bands and equipment separate.

Present-day communication performance to ships is very bad
with the average delays in getting a message of eight to ten hours
because of erratic HF propagation; congestion and interference in
the relatively narrow bands allocated; and non-technical problems
in the area of labour/management, tradition and economics.

Technical progress in the maritime satellite field has been
more successful thus far than the organizational and political
progress.   In early 1972 COMSAT successfully demonstrated two-way
transmissions between an experimental station on board the "Queen
Elizabeth II" and a small earth station at COMSAT's Lab. in
Clarksburg, Maryland.   The weather varied from 90° F to a snow
storm; tests included transmissions of voice. facsimile, teletype,
biomedical data and computer access, all successfully accomplished.
While the equipment operated at 4/6 GHz to permit testing via the
Intelsat IV satellite, much useful information was obtained concerning shipboard equipment characteristics for this application and
this information has been used in drawing up specifications for
the first generation of operational equipment.   Additional testing
was conducted at VHF frequencies by engineers from EXXON Corp., and
General Electric using NASA's ATS-1 and ATS-3 satellites.   Here,
too, successful tests were performed with various types of transmission, and in addition automatic ranging of the ship when using
both satellites simultaneously indicated that in a future VHF system
it is reasonable to expect positioning fixes to an accuracy of one
nautical mile.   It is possible to achieve an order of magnitude
improvement, to a tenth of a nautical mile, by going to the L-Band
system at 1500 MHz.

4.2 Present Status.   At the present time there are two maritime
satellite projects underway.   In addition, international negotiations have been initiated leading to the possibility of creating an
international organization for maritime satellite communications.

4.2.1 MARISAT.   This is a joint venture of the Communications
Satellite Corporation (86%), and three other U.S. record carriers
(14%), to put two satellites in orbit during the end of 1975 or
early 1976.   These satellites - one at 176.5° E longitude, and the
other at 15° W longitude - will initially have the bulk of their
capacity leased to the U.S. Navy for communications in the lower
end of the UHF band.   In addition to this equipment, the satellite
(Fig. 3) will include communications repeaters operating in the
commercial maritime frequency bands around 1600 MHz.   Since the
Navy requirement is expected to last for only a couple of years,
and its use during that time will call for differing amounts of
satellite power, a very flexible design has been chosen for the
commercial maritime application.

COMSAT is purchasing two-hundred shipboard terminals for this
project.   The automatically steered antenna (Fig. 4) is four feet

Figure 3 - MARISAT Satellite

Figure 4 - Antenna of Shipboard Terminal for MARISAT

Figure 5 - Below Decks Equipment of Shipboard Terminal for MARISAT

in diameter (1.25 m) and is housed in a radome for protection from ice and snow.   Its 10° beam will permit a relatively simple auto-steering system to be used.   All equipment is completely solid state even including the 40-watt transmitter output stage.   Great attention has been paid to simplify the operation of the equipment and to provide long, trouble-free life under maritime conditions.

The channel capacity of the system, with the various power levels discussed above, is as follows:

Power Level	Low	Medium	High
Telephone circuits	1	5	9
Plus teletype circuits	44	88	110

A view of the antenna in its protective radome is shown in an actual shipboard installation (Fig. 6).

4.2.2 <u>MAROTS</u>.   This is a maritime satellite being developed by the European Space Agency for a number of European countries, with Great Britain, the major partner, financing over half of the project, and eight other countries underwriting the remainder.

The project was started in 1973 and is scheduled for launch by a 3914 Delta rocket in October 1977.

It is estimated that the project will cost about $90 million (mid-1973 price level) and will result in the launch of one "pre-operational" satellite towards the end of 1977.   It will be positioned at about 40° E, so that together with the two MARISAT satellites, substantially global coverage will be provided.   Since this satellite will be launched about two years after the MARISAT, it was designed for launch by the 3914 Delta rather than the earlier 2914 Delta used by MARISAT.   This will provide several hundred pounds greater payload for MAROTS, which, together with its three-axis stabilization will provide about four dozen telephone circuits compared to the dozen of the MARISAT.

4.2.3 <u>INMARSAT</u>.   There have been several meetings called by IMCO to form an International Maritime Satellite Organization (INMARSAT), which will be a global maritime satellite communications system. At its plenary conference in London, April 23 to May 9, 1975, agreement could not be reached on four major subjects and an inter-sessional working group was set up to try and resolve these issues before the next plenary on February 9 to 27, 1976, in London.

4.3 <u>Future Possibilities</u>.   While MARISAT and MAROTS, as with any satellites of new design, may exhibit some initial difficulties there is no question of the technical soundness of satellite communications for maritime use.   Just as the last decade has seen the

Figure 6 – Radome Installation Aboard Ship

growth of fixed service international satellite communications from
an experiment to the major means of transoceanic communications, it
can be confidently anticipated that the next decade will see maritime
communications swing over from HF to satellites.

Even before the launch of MARISAT and MAROTS, plans are being
formulated for more advanced designs providing more channels per
satellite and therefore a lower cost per channel. The basic
approach here, as with the future aeronautical satellite, is to
employ a multiplicity of narrow beams illuminating an ocean area
thus obtaining higher effective radiated power per beam and offering
the potential to re-use the frequency several times in beams
sufficiently separated to avoid interference.

5.   BROADCAST   SATELLITES

5.1 ATS-6 Broadcast Satellite, U.S./India.   The first experimental
satellite designed for transmitting TV directly to a small diameter
antenna (2-3 m) was the ATS-6 (Fig. 7) launched on June 1, 1974, and
used for one year over the U.S.   The unfurlable satellite antenna
of 9.25 m diameter and an associated 15-watt transmitter at approximately 2600 MHz provided a useful signal for all stations lying
within a circle of roughly 650 kilometer diameter.   Transmissions
were made to regions in the Rocky Mountains and also the Appalachian
Mountain areas for teacher-aid programmes; and for health care in
the U.S. northwest, and Alaska.   The groups involved were enthusiastic about the project but were unable to raise funds for a
second satellite of this type.

Operation over India started in August 1975 at 860 MHz on a
beam which covers India.   The Indian government is locally building
about two-thousand-four-hundred earth stations with 3 m diameter
antennas (costing about $1,000 plus $300 for the TV receiver) for
use in villages for reception of the satellite signals. Programmes
have been written for a three-month period calling for transmissions
four hours/day of which roughly half is instructional material for
children, and half for adults including material on literacy, hygiene,
birth control and agriculture.   Thus far, the Indian government has
spent about $20 million for the project - about half for the three
months of programming and the other half for the stations, including
several transmitting stations; re-broadcast stations; and the two-
thousand-four-hundred small receive-only stations.   By the summer
of 1976 the satellite is scheduled to be moved back to the U.S. and
there are as yet no plans to replace it.   This will be the first
full-scale demonstration of broadcast TV direct to villages for
elementary educational programmes and the future course of similar
TV broadcast projects may well depend on the results of this
programme.

Figure 7 - ATS-6 Satellite

5.2 Canadian Communication Technology Satellite (CTS). This
satellite is a joint Canadian/U.S. venture in which Canada will
design and build the satellite (Fig. 8) while the U.S. will furnish
the specially-designed high-power (200 watts) TWT operating at
12 GHz; and also the Delta launch vehicle. The purposes of this
experimental satellite are to demonstrate TV transmission at this
frequency to small earth stations using antennas 2.4 m in diameter
for TV and 0.9 m diameter for FM sound broadcast; and to develop
and flight-test spacecraft subsystems and components for use in
future communications satellites. To operate with antennas of
this small size it is necessary to restrict the satellite antenna
beamwidth to 2.5° x 2.5° which will therefore cover only a fraction
of Canada at any one time. The satellite is to be launched in late
1975 or early 1976.

5.3 Japanese TV Broadcast Satellite. The General Electric Company
is building an experimental satellite for broadcast purposes for
Japan. This satellite will transmit two TV channels both in the
12 GHz band - one channel will connect to an antenna illuminating
the main islands of Japan while the other channel will illuminate
an area about ten times as large and include islands away from the
main islands as well as those also. For the main islands the earth
station receiving antennas will have a diameter of 1 to 1.6 m while
for the second channel an antenna of 2.5 to 4.5 m will be needed
depending on the location.

5.4 U.S. Programmes. There have been no specific proposals made
within the U.S. for a 12 GHz satellite designed to transmit directly
to the home because ninety-seven per cent of U.S. households already
have TV sets with access to at least two channels and of these
households ninety-eight per cent have access to three or four
channels.

However, there are two projects underway for satellite transmission of programmes for rebroadcasting purposes*. In one case
it is planned to pick up programmes prepared for CATV use in one
part of the country and to relay these programmes via the satellite
for pickup and use by CATV systems in other parts of the country.
It is estimated that a receiving station with a 10 m diameter antenna
and associated equipment can be built to sell for $75,000 to $100,000.

The Public Broadcasting Service is also considering leasing one
or more transponders from domestic satellite systems to transmit
their programmes throughout the U.S. to one-hundred-and-fifty earth
stations employing 8 m antennas from which they will rebroadcast
over existing PBS TV broadcast stations.

---

*Both projects use the 4/6 GHz bands.

Figure 8 - Communications Technology Satellite

It is pointed out that transmission costs in the U.S. tend to be a small part of total cost; for example, the three major U.S. TV networks' annual bill for transmitting their programmes around the country is only about six per cent of their gross.

5.5 Summary.   While there is little doubt as to the technical feasibility of the above-mentioned projects, there is a question as to the extent to which the broadcast satellite will either replace or augment present systems for education and entertainment. Within the next couple of years the programmes mentioned above should provide an answer to this question.

6.   DOMESTIC SATELLITE SYSTEMS

6.1 General.   Satellites offer a potential for establishing rapidly and simply a domestic communications network for countries where difficult geographical features, size and state of economic development have heretofore inhibited the development of such a system.   They also offer in the case of countries with a developed network an additional means for communication flexibility and diversity together with the possibility of providing new and different kinds of communications services not easily provided otherwise.

6.2 MOLNIYA/STATSIONAR - U.S.S.R.   The first domestic satellite communications system was the U.S.S.R.'s MOLNIYA-1 which started operations in 1965 using frequencies around 900 MHz.   It used an elliptical orbit with a twelve-hour period, an apogee of forty-thousand kilometers, a perigee of five-hundred kilometers, and an inclination of 65°.   This orbit is well suited for transmission to northerly latitudes as needed for the U.S.S.R.   Three satellites are needed in orbit simultaneously to provide full twenty-four-hour coverage.   While the MOLNIYA-1 has been augmented with a MOLNIYA 2/3 series in a similar orbit but operating in the 3.5/6 GHz region, plans have also been put forth for synchronous satellites operating in the same frequency band.   At this time the number of earth stations has increased from two dozen to about twice that number.

6.3 ANIK - Canada.   In early 1973 the Canadian TELESAT system became operational using the ANIK satellite (Fig. 9).   Some fifty antennas were used, the majority of them incorporating 7.2 m diameter antennas which are used for TV reception and/or telephony with one or two voice channels per station.   These stations are located in remote areas which previously depended only on HF communications. In addition there are about a half dozen stations with 9.2 m antennas which are used for TV distribution and a couple of stations with 30 m antennas with a capability to handle nine-hundred-and-sixty telephone circuits between Toronto and Vancouver.

Figure 9 - ANIK/Western Union/Indonesia Satellite

6.4 <u>WESTAR/Western Union - U.S.A.</u>   In 1974 the Western Union Co.
of the U.S. initiated a domestic satellite system using the same
type of satellite as the Canadian ANIK but with an antenna beam
modified to cover the U.S.   This system includes about a half
dozen earth stations with 15.5 m diameter antennas located near
the major cities in the country.

6.5 <u>COMSTAR - U.S.A.</u>   COMSAT General Corp., a wholly-owned sub-
sidiary of COMSAT, has a contract to develop, build, launch and
maintain in orbit three satellites which will be leased to AT & T
for domestic use in its national telephone network.

Unlike the ANIK design which is launched on a Delta vehicle,
the larger COMSTAR satellite will be launched on an Atlas-Centaur.
The COMSTAR satellite (Fig. 10) is the first satellite proposing to
use two polarizations, wherein twelve repeaters connect to the
antenna with one sense of polarization, say vertical, and another
twelve connect to an antenna with an opposite sense of polarization.
The twenty-four repeaters on a satellite have a total capacity of
twenty-four one-way colour TV channels; or fourteen-thousand-four-
hundred two-way telephone circuits; or a digital capacity of
approximately one billion bits/sec., or any combination of these.
AT & T will own and operate five earth stations using 30 m diameter
antennnas.   The first satellite of this series is to be launched
in the first half of 1976.

6.6 <u>Indonesia.</u>   Construction is underway for a domestic satellite
system for Indonesia.   The satellite is similar to the ANIK and
forty earth stations are being built for operation in this system.
These will use 7.3 m or 9.2 m diameter antennas and will provide
telephony and TV service.

6.7 <u>Aetna/COMSAT/IBM - U.S.A.</u>   In conformance with an FCC deter-
mination on eligibility critera, a joint venture has been formed
between Aetna, IBM and COMSAT General to establish and operate a
domestic communications satellite system.   The system will be
designed to operate at 12/14 GHz using 5 m diameter antennas
located at the customers' premises.   It will be an all-digital
system designed to provide a broad range of communication services
to major industrial users.

6.8 <u>Other Systems.</u>   A number of countries are leasing, or planning
to lease, repeaters (sometimes called transponders) from INTELSAT
to use for their own systems.

Spain and Mexico have a half transponder which is used for TV,
mainly from Spain to Mexico.   Norway has leased a half transponder
which will be used to provide communications to oil rigs in the
North Sea.   Algeria has a dozen stations operating through one
transponder for communications from its northern cities to its

Figure 10 - COMSTAR Satellite

southern towns a thousand miles away across the desert.   Brazil
has one transponder for communications among three stations
separated by jungle.   Other countries planning to lease transponders include (with the number of transponders planned) Zaire (1);
Colombia (1/4);   Nigeria (2);   and Malaysia (1/2).

Iran and the Arab countries are actively engaged in planning
their own satellite systems.   Japan has awarded a contract to
Aeronutronic Ford Corporation for a satellite for domestic use,
using both 4/6 and 20/30 GHz frequency bands.   This is to be
launched in 1977.

7.   IMPACT   OF   COMMUNICATIONS   ON   THE   WORLD   OF   THE   FUTURE

The last decade has seen an extremely rapid growth of two technologies - one of digital techniques as used in computers;   data
processing;   and in digital voice coding;   and second, the growth
of long distance transmission, especially via satellite.   Some of
these technologies which are relevant to this discussion are
summarized below.   An attempt is then made to envisage systems
which might be obtained by combining developments in these two
fields and some observations are made on the possible impact of
the resulting systems on future society.

7.1 Digital Devices, Techniques and Related Equipment.   Large
scale integration with over a thousand transistors per four millimeter chip.

Adaptive facsimile systems which in effect send only the
information content of a page and ignore large black-and-white
areas.   Such systems have been built and provide an order of
magnitude of decrease in the time for scanning a page.   High
speed facsimile machines which scan one page, with excellent
resolution, in one to two seconds are under development.

Automatically equalizing modems so that data transmission via
ordinary telephone lines is increased from twelve-hundred bits/sec.
to four-thousand-eight-hundred bits/sec.;   and to nine-thousand-
six-hundred bits/sec. on conditioned lines.

Constantly decreasing memory circuit costs which are down to
roughly one-tenth cent per bit in large production quantities with
further promise of a twofold reduction in price in a few years.

The widespread use of push-button telephone sets providing a
digital input to the telephone system in the home or business office.

The intensive development of low cost terminals aimed at cutting
down the cost to an amount where such terminals can be realistically

considered for use in all business offices and even in the home.

7.2 Growth of Communications.   The last decade has seen the
capacity of individual satellites increase by a factor of twenty-
five times.   (Intelsat-1 with two-hundred-and-forty circuits, 1965
launch, compared to the six-thousand-five-hundred circuits for the
Intelsat IV-A, launched in September 1975.)

The satellites which COMSAT will launch for AT & T for U.S.
domestic use will each have a capacity of over one billion bits/sec.

All commercial international communications satellites built to
date and also the domestic satellite just mentioned above are all in
the band 4/6 GHz.   With the opening of the 20/30 GHz band, its
greater bandwidth and the use of multiple beams could provide a
digital capacity of ten times the above, or ten billion bits per
second.   Terrestrial systems for cable, microwave relay, and wave-
guide are being planned for digital rates of two-hundred-and-seventy-
four megabits per second per channel, which could readily interface
with future satellites.

While the above describes the large, high capacity digital
systems which will be used in the future, a substantial amount of
digital circuitry already exists in telephone systems, and work is
underway on expanding this approach further.   For example, AT & T
has over two-million channels of the T-1 twenty-four-telephone
channel digital multiplex in operation (1.5 megabits/sec.) and with
the introduction of a new lower loss cable, two such multiplexers
can be combined to send forty-eight digital telephone channels over
the cable; which can also be used for 6.3 mbps transmission.   By
1990 the telephone plant is expected to be predominantly digital.

The development of optical fibers has also progressed during
the last decade from a laboratory curiosity, with losses of hundreds
of decibels per kilometer (quite impractical for operational use) to
its present loss of a few decibels per kilometer which is so low
that active work is underway at a number of laboratories around the
world to apply these fibers to long haul and short haul transmission
systems, and eventually even to the home, to provide a wideband
facility with megacycles of bandwidth and without the requirement
for amplifiers between the home and the nearest telephone office.

The purpose of discussing non-satellite digital transmission
systems has been to bring out that the large digital capacity of
satellites is compatible with the very substantial effort on digital
transmission for terrestrial systems, which in many cases will be
needed for the interface with the satellite system.

While satellites have the capability to span distances of
thousands of miles with tens, or even hundreds of thousands of

voice circuits;   there exist other needs for which the satellite is
well suited.   A number of industrial plants and offices of a single
corporation have need to communicate information from one site to
one or more selected other sites, or to all simultaneously in a
broadcast mode.   The nature of the information will vary during
the day but may include voice, teletype, data and facsimile signals.
For economic and physical reasons it is desirable to keep the
antenna diameter to say 3 m to 5 m so that it can be installed on
the rooftop of a building.   The use of the 12/14 GHz band is well
suited to this application since this band is not shared with
terrestrial microwave services.   Earth stations can therefore be
installed in a city without fear of interference with terrestrial
microwave facilities.   Also, since there are no flux density
limits, increased satellite power can be used to result in smaller
diameter reflectors on the ground.

Since any one station would use only a small fraction of the
satellite power, hundreds or thousands of such earth stations will
be needed to fully utilize the satellite capacity, and therefore it
is essential that these stations be relatively low cost, and therefore unmanned.

7.3   Possible Future Systems

7.3.1   Computers, Terminals, and Communication in Education.   Much
has been written on this subject and studies and actual experimentation are underway.   The range of automation in teaching is very
wide, encompassing the following.

Technical possibilities range from students at a neighbourhood
learning centre seeing and hearing a remote lecturer via a closed
circuit TV system with an audio return for questions;   to a terminal
in the student's home to permit him to dial into the school for any
course he wishes to see and hear.   Similarly, computer-aided
teaching systems can be made available at remote terminals either
in the school itself, or at neighbourhood centres, or in the home.
Consideration of such systems raises serious questions as to the
relative merits of the educational benefits obtained from these
methods compared to the present system;   the relative economics;
and the effect on the teaching community.   These questions can
best be answered by substantial pilot programmes.

7.3.2   New Business Systems.   We are already using electric type-
writers with memory, which permit making corrections and additions
to a text without the need for the typist to re-type the entire
manuscript.   The memory bank can be enlarged and used to automatically file all letters and documents typed into that machine.
These can be indexed and cross-filed under various headings, to be
available for instant recall.

The concept can be expanded to permit not only the instantaneous filing of a memo, but simultaneously its transmission to a remote office where a number of copies are automatically prepared and addressed to all persons in that office who have previously expressed an interest in that subject and requested the receipt of copies of related memos.   Only one "copy" need be filed for an entire plant.

By means of digital techniques an office's communications needs, i.e. telephone, telegraph, data, facsimile and videophone, will all be handled through a digital bit stream and transmitted via a satellite to selected other offices or to all on a broadcast mode. Great flexibility is possible since the output bit stream can be made up of different mixes of the several types of signals and these can be constantly varied during the day.

7.3.3 The Home Communications Centre.   It has been proposed that each home have a wideband link to the local central office rather than the 4 KHz twisted pair as at present and that the home terminal be used to send and receive mail; for reception of the daily news- paper by facsimile; as a students's "classroom" during the day; for the housewife shopping; as a videophone; and for a directory of the town's activities.   Technical means exist to permit accomplishing all of these functions, but aside from the technical problems there are significant economic and social questions to be answered.

7.3.4 Automatic Translation.   Making available high quality tele- phone circuits between one-hundred different countries highlights the problem that many of these speak different languages.   In this age of the computer, is it feasible to think of a machine into which one language is spoken and out of which the same sentence is produced in a different language?

The problem can be broken into two parts; that of speech recognition, and that of translation.   With research equipment presently available, about eighty per cent of the words are recognized correctly, but a rather large computer is kept busy at this task.   However, the project is in its early stage and will undoubtedly be improved in the future.

The second problem, of machine translation, is more difficult. A number of groups working in this field are now emphasizing the fundamentals of linguistics as a first step in the ultimate goal of achieving a translation machine.

Eventually, the hardware of the speech recognition groups, coupled with the findings of the linguistics groups, should provide the background to attack the problem of direct speech translation.

The availability of a computer translation facility, plus the global circuits of the satellite system, should together provide not only worldwide technical communications, but hopefully better worldwide understanding and real communications.

7.3.5 <u>Conclusions</u>.   The previous sections have discussed the applications of electronic computers and communications to education; to business, and to the home.   In most cases the technology already exists to implement such systems, but there are serious questions of economic viability and social acceptance.   As our technological capabilities expand, the new systems which are then made possible also expand and have an even greater impact on society.   There is developing an increase in concern by engineers for the non-technical aspects of their work.   Recently, a special issue (about 175 pages) of the IEEE* Transactions on Communications, October 1975, Volume COM-23, Number 10, was devoted entirely to "Social Implications of Telecommunications".

Professor M. Schwartz of Columbia University, in a guest editorial**, summed up the issue by noting that:

> "No one obviously knows which way the technology will move us. We do not know whether the problems raised above in the area of citizen-government interaction will materialize, or whether the net effect will be a positive one.   We do not know whether communication alternatives to transportation, as another example, will lead to boredom and decreased person-to-person interaction, in addition to the desired effects of decreased energy use, reduced travel time, etc.   Changes in our modes of communication can lead to a more homogeneous society or to a more heterogeneous one.   The employment skills required with our information revolution, calling presumably for more white collar and better educated employees, may further discourage the current 'have nots'; on the other hand the improved educational opportunities made possible may decrease the gap between the 'haves' and 'have nots'.   No one knows – but raising these questions may lead to their solution.   Noting the possible problems may suggest alternative approaches to minimize possible problem areas."

Several authors stress the great difficulties involved in "sociotechnical" research especially to engineers who are accustomed to physical research, viz. – the lack of fundamental laws such as those which they are accustomed to in engineering; the impossibility of isolating the problem so that it can be attacked in a laboratory; the interactions of people with their complex motivations; the need to test for months or years before valid conclusions can be drawn.

---

* Institute of Electrical & Electronic Engineers
** ibid, p. 1009

Gillette* states that "the most important measure of success for a novel application is likely not to be an abstract cost/benefit calculation, but rather the vital market place evaluation of whether anyone wants to support an operational version of an experimental system".   He quotes Barnett** who believes that the following questions must be answered positively in evaluating a new medical data system (but they apply equally well to other systems):

a) "Is the system in daily operational use meeting real needs?

b) Is the system supported from the routine operating budget?

c) Has the system been transferred and used elsewhere both by other medical institutions and private industry?"

In order to answer the "bottom line" criteria of Gillette and Barnett it is first necessary to implement a pilot installation of sufficient extent and duration to provide meaningful results. Even pilot installations of interactive CATV systems; computer aided educational systems; electronic mail systems; and advanced electronic business installations; would each require investments of millions or tens of millions of dollars over a period of several years before the above questions can begin to be answered.   However, such costs are only a small fraction of what a full scale national system would cost, and because at present we have no social calculus to permit a theoretical analysis of these complex socio/technical/economic systems, we cannot avoid the need for pilot installations.   There is no royal road.

---

* ibid - "Using Discretionary Telecommunications" D. Gillette, P 1054
** "Medical Information Systems at the Massachusetts General Hospital"
G. O. Barnett, Proc. Int'l Conf. Health Technology Systems,
M. F. Collens, ed. Operations Research Society of America, 1973,
San Francisco, California

THE IMPACT OF SPACE COMMUNICATION

G. Rosenberg, Norway

An edited summary of the discussion

1.   INTRODUCTION

Communication satellites are one of the most important applications of space science.   Exploiting wisely the future possibilities
will help us to overcome problems between men and nations due to
lack of communications.   But it is the problem of our generation
to make adequate preparations in order to avoid traumatic experiences
in the future.   We have the responsibility to understand the technological development and take the necessary actions in planning and
decision-making.   The techno-economic maximization is only part of
the solution.   More important are the administrative and political
implications.

Some caution is needed however in the process of harmonizing
the new technology with the existing in order to avoid a "tail wags
dog" situation.   Telecommunications already have a long tradition
as an earthbound service and have gone through a tremendous development since the days of Bell and Marconi.   Industry and organizations
have developed and international bodies have, with considerable
success, regulated conflicting interests.

Space communications is one step in this development, but gives
new dimensions which concern us all.   The challenge here is to
establish the proper environment for the full exploitation of satellites.   If they are constrained by the framework of earlier communications technologies the full benefit will not be received.
If, on the other hand, they are developed without reference to
earlier communications technologies they would also not be able to
evolve successfully.   Considering the long-term technological

prospects and their technical, administrative and political consequences we must try to build the future on the vigorous parts of
the past.

2. LONG TERM TECHNOLOGICAL PROSPECTS

The general impression is that the technological surge is still
continuing and that, with some exaggeration, everything is still
possible.   This is especially true for the United States and to
some extent for Japan, the Soviet Union and several other industrialized countries which have advanced their capabilities significantly in recent years.   The main issues are therefore not technological, but related to the international technological competition
and to the gap in technology between these countries and the rest of
the world.   These differences in level of technology supports fears
of economic and political dependence in many parts of the world.

A typical European point of view is that today's monopoly of
technological development poses a problem.   Either Europe accepts
the monopoly and tags along behind, or European industry tries to
participate, a choice which inevitably will lead to commercial con-
flict.   Are we in this situation going to share technological
information or proceed in parallel, wasting large amounts of money?
There is a growing paradox in the acceptance of the need for better
communications and the desire to share the profit by participating
technologically.

It was also pointed out that the same monopoly problem exists
in the management of international communication satellite systems
and in the present launcher situation.   The latter situation will
change when the European Space Agency develops a launching capability in the 1980's.

With the existence of many domestic and regional communication
satellite systems and possibly also several international systems,
what are the prospects for tracking and data relay systems for
intersatellite communications?   In the United States NASA has two
studies underway as a result of which specifications for such a
system will be developed.   The hardware development will eventually
be made by private companies.

Why do we have independent satellite systems for different
purposes such as communications, meteorology or earth resources?
How can we share systems and earth facilities?   No unique answer
exists.   Some systems can be combined, some cannot.   Several
difficulties appear:

    a) Different technologies have to proceed at the same pace
       in order to retain flexibility.

b) Combination of different instruments and co-operation of different organizations.

c) Political problems, tradition and prestige.

d) Economic problems such as billing and sharing of expenses.

Typically, the combination of aeronautical and maritime services which is advantageous from a technical and economic point of view, has been impeded by political and organizational difficulties.

In the future it will be possible with the tracking and data relay system to use various sensor satellite systems, both low and synchronous orbit, together with, for example, the Intelsat communication system. The system presently under development by NASA has a more limited aim, which is to replace NASA's twelve to fifteen tracking stations all over the world with one single earth station in New Mexico.

The gain in co-operation today is more likely to be found in the standardization of sub-systems.

In the near future the communication capacity will increase by a factor five to ten and satellite lifetime will increase to seven to ten years. Reasonable power levels for direct TV transmission can already be reached.

Twenty years ahead we may envisage multipurpose manned space stations servicing unmanned stations, updating equipment, making replacements, repairing, etc.; and instead of paying according to traffic, customers will rather be billed for number of ports.

Launcher availability is not a technological question, but a cost/efficiency problem and a political problem. Indeed launcher costs have increased substantially lately and the Space Shuttle seems to be the potential low cost system of the future. The co-ordination of payloads, however, places undesirable constraints on the technological development. This again influences cost since the number of launches will depend upon usage.

Perhaps somewhat unrealistic plans have been made for sixty launches per year, making operation and usage similar to that of airlines and making repair, replacements and other services feasible at least from an availability point of view. Approaching the other extreme with two launches per year or less, the whole situation will change drastically, making traditional launchers more competitive and putting more emphasis on redundant constructions and multi-satellite systems with switching for increased capacity and alternative routing.

A service capability is necessary, perhaps not so much for communication satellites which have reached a more mature technological level, as for the sensor systems. This capability will at first only be established for low orbit systems but a definite need exists for eventually servicing synchronous satellites as well.

The trend for the future is towards smaller and more numerous ground stations and more sophisticated satellites. This trend in information distribution has huge social and political consequences and the scientific and technical community have an obligation and a responsibility to try to show what the impact will be on the individual, on society and on all other parts of life. This amounts to far more than marketing and should not be left to the space agencies alone, but should also be done by the PTT administrations which in this context are closer to our social and economic conscience. It should be noted, however, that the introduction of satellite capabilities could have an adverse effect on investment in conventional terrestrial facilities and as such, the PTT administrations may not have the desire or motivation to introduce the satellite possibility even though economic and service benefits may be substantial.

Another and quite different aspect of the future is the change in industries as physical communications decline and the electronics industry prospers. A new equilibrium will be found.

3.   COMMUNICATIONS   TECHNOLOGY   FOR   MOBILE   SERVICES

Satellite communication is in the process of rather dramatically changing available mobile services. Increased communication capacity and improved navigational accuracy permits large scale remote control with a corresponding shift in responsibilities.

There are today fifty-thousand ships larger than one-hundred tons. This may serve as a background for a human note. The captain's autonomy has a long tradition in maritime history and will influence greatly the introduction of maritime communications. This raises the question of how much communication and navigational aids are really necessary.

For locating and interrogating buoys and drifting platforms satellites are obviously important. What is useful information on board a ship depends upon the degree of customer orientation in content and form of presentation. Investigations show that administrative traffic will be low, perhaps a teletype channel a few hours per day, but that there will be an increasing demand for welfare traffic such as crew-home communication and radio and TV contact with home country. This may be envisaged for aircraft too.

Modern management has affected the captain's autonomy and made sailors adapt very much already.   Developments in computer techno-logy, however, has led to a trend for as much computer power on board the ship as possible.   This tends to balance the ship-shore relationship.   Socio-technological studies are necessary in order to find the right sharing of responsibilities.

There is a need today to set up managing organizations for the different maritime services in order to share the expenses.   It is highly desirable to prevent a multiplicity of systems and have a co-operation, say, between the American and European maritime satel-lite communication systems MARISAT and MAROTS, ultimately leading to one organization.   The INMARSAT conference organized by the Inter-governmental Maritime Consultative Organization, IMCO, is proceeding towards its objective of creating a single organization for maritime satellite communications.

An important application of satellites for mobile services is for search and rescue operations where a centralized system could be more effective in filtering out false alarms and give better alert-ness.   Several concepts for such systems have been studied but have not been well received.   The problem seems to be psychological rather than technical.

One aspect of maritime services which has been very well received is the navigational satellite system which for some appli-cations, such as for example ocean current studies, has led to a revolution in navigation.   For most applications, however, the present Transit System is too accurate at open sea and not good enough in confined waters.   In the early 1980's the Global Positioning System will provide a 20 m accuracy with one-hundred per cent accessibility.

4.   GOVERNMENT   ADMINISTRATION   AND   INDUSTRIAL   MANAGEMENT

Many large-scale effects are already observed.   Direct dialling has made censorship very difficult.   The role of the ambassador has changed through direct government contact and the hot line has had, and will have, an effect on world affairs and on the organization of governments.   The same effect is seen on military operations and organizations.

The advances in communications can lead to increased centraliza-tion both in industry and government, but can also be consciously used for decentralization and can reverse the trend towards increased urbanization.   The tool is neutral.   Information must be fed both ways in the pyramid leading to more decisions on a lower level.   The problem is similar to that of the autonomy of the captain and is really a sociological problem which has to be solved as such.

Advanced communications may also save resources and make time more efficient by reducing personal travel to business meetings and by giving more time to meaningful social contacts and a better way of life.   The danger, however, lies in increased alienation through reduced person-to-person contact.   A compromise must be found. Telecommunications are necessary but not sufficient.   The point is that we will be given the choice between business meetings via satellite or personal travel.

In many fields centralized large data files connected through data networks become necessary, although the risk of damage by accident or fire will lead in the direction of having the same information stored in more than one place.   Since it will be possible through satellites to transfer and update data bases, redundancy and flexibility will be achieved.

Industrial applications include computer-aided design, marketing and economics while social services is a typical government application.   Here computer security or access control becomes a problem.

Abuse of data libraries is a threat against privacy, human dignity and responsibility.   Things happen piecemeal and politicians do not see on the right time scale.   We have to educate them.   Who else?

Good communications are a prerequisite for efficient operation of both government and industrial enterprises.   Integrated economic models for day-to-day planning become possible.   But there must be an emphasis on what management needs rather than overselling what can be given at any moment.

We must learn to control the information explosion.   What we want to achieve must be matched to what is possible.   In this context we want to improve decision-making.   The problem here is not data communications, but data processing and presentation because our brain capacity is not increasing in the same proportion as the flow of information.   Enormous amounts of information cannot be assimilated because our thinking time has become the bottleneck.

5.   INFORMATION   DISTRIBUTION   AND   EDUCATION

Instant information distribution and direct educational television will have an enormous impact on our culture.   Some of the problems are outside the scope of technology but are at least as important to solve.   And even though technology is neutral, technologists are not.

Adult education will benefit most from educational TV.   Several experiments are in progress or are being planned, the most

interesting being the Indian utilization of ATS-6.   Around three-
thousand $500 receiving stations are being set up in Indian villages,
covering 2.5 per cent of India's villages but reaching thirty per
cent of all teachers.   With programmes in four languages on hygiene,
family planning, national integration, news and entertainment, the
impact is tremendous.   There are villages which hardly know that
the British have left.   Two-hundred sociologists are involved in
investigating viewer responses, evaluating programmes and making
new programmes.   One of the lessons learned is the necessity to
change programmes faster than anticipated, but it is far too early
to judge the full impact.   However, as Indian plans extend only
for one year, major effects may never be observed and the discon-
tinuity may create undesirable after-effects.

Brazil used ATS-6 for educational programmes in a similar way.
But education is not a passive process.   This was demonstrated in
the UK with the failure of The University of the Air programme.
After redesigning it is now a success called The Open University
with education proceeding mainly through correspondence and TV just
giving the final touch.

Thus a low cost two-way system is desired.   The modest information flow in the reverse direction makes the problem easier.
Again using ATS-6, a successful experiment in interactive teaching
was performed in the Rocky Mountains and Alaska.   This could possibly reverse the unfortunate trend of bringing people to the cities
for educational purposes; it counteracts the lack of teachers
especially in developing countries, it can be used for continuous
education of teachers and doctors and for real time consultations
in emergencies.

The United States, however, is in the deplorable situation of
having discontinued the educational programme due to lack of support.
This is a marketing problem.   The most fertile ground for applications are in the developing countries and perhaps in Europe where
education and health services to a larger extent are government responsibility.   This certainly raises the important questions.   Who
makes programmes?   Who is responsible?   Who is in control?

Further experiments in educational TV will be performed by
Germany and France in 1976 using the Symphonie 1 and 2 satellites
for a number of African countries.   It is important that an exchange of information on the experiences from all these experiments
takes place.

6.   THE   FUTURE   LIBRARY

It has been estimated that the information doubling time in
1945 was twenty-five years, in 1955 twelve years, in 1965 five years

and in 1975 three years.   Knowing the enormous cost involved in
building and keeping new libraries up to date, libraries which look
very much the same all over the world, it seems that the time is
ripe for computerized libraries.   One world-wide system is technically feasible from a storage and accessibility point of view, but
the large problem is that of classification and retrieval.   Cul-
tural differences create barriers which are difficult to surmount.
Computers with associative memories may in the future be the solution to this problem.

In various disciplines there are world data centres already
working, but these are generally solutions to far simpler problems.
The interactive ESA system which provides US data banks in Europe
is a good example which should be combined with the Common Market
efforts to provide one European system.

As the costs of sorting out outdated literature is higher than
storing it, one could be tempted to suggest throwing away everything
older than, say, five to ten years.   With today's information explosion, not very much would be lost.   But this is implicity also
the objection.   Not very much would be gained either.   Then, as
was retorted, one should rather burn the last five years' literature,
forcing scientists to think five years before writing.

After this exercise in futility it is rather more important to
suggest adaptive interactive systems where indexing can change with
time.   To fulfil all needs rather advanced interactive systems are
required with displays and high-speed hard copying devices and some-
body at the library end of the system.

However, science does not depend critically on good documenta-
tion.   And strong reservations are taken against possible governmental use of quotation rankings as an aid in science support, an
idea which has been suggested to challenge the peer system.   Science
works on individual decisions.

And what about automatic translation in connection with the
future library?   The idea did not stir up much enthusiasm.   Rather
teach people the few necessary languages, in the future perhaps only
two or three.

7.   STABILIZING   AND   DESTABILIZING   EFFECTS   OF
INCREASED   COMMUNICATION

It is already a well-known fact that mass distribution of information can create effects of major importance on a national or
even international scale.   Direct TV-transmission from satellites
will enhance this process and will influence national and regional

attitudes and policies as well as local emotions.   In what direction?   The Damocles sword is easily seen.

In Canada the domestic satellite project has as one of its
principal aims to tie the Eastern and Western Provinces closer together.   TV-coverage, although limited, had a large effect on the
termination of the Vietnam war.

But the problem is really extremely complicated and the impact
difficult to trace.   Nevertheless it is one of the important problems we face today.

A minor strike instantly transmitted into each home can create
similar actions other places, while the opposite, no transfer of
information, leads to a charging effect with not very well-known
consequences.   With the two related major problems facing the
world today in mind, overpopulation and food production, how will
instantly available information on famine and affluence shake our
equilibrium?

The view is generally optimistic.   Information will improve
the world.   It has been said: "Know the truth and the truth will
make you free".   Global communications are the precursors for a
world federation.   Just as the telegraph and the railroad were
necessary ingredients in the creation of the United States, so will
satellites and jet planes have the same impact on the world.   The
danger lies in biased or unbalanced presentations, manipulation of
news and propaganda; sensible news not highlighted, riots exag-
gerated.   Therefore many voices should be heard through a pluralis-
tic system.   Better and broader information is a progress and cannot be blocked in the long run.

But should only one system exist?   A regional system for
Europe is desirable for the unifying process.

Furthermore, as systems get so complicated that only a few
people can handle them, much power and large responsibility are in
the hands of few.

The Intelsat system is available for all.   Many systems would
be impractical for the smaller countries.   Therefore an international joint venture is the only economic solution.   This does
not prevent regional or national systems, but an international system must have first priority.   With another approach the progress
that has been made would not have been seen.   Intelsat was
initially US-dominated but as technical capabilities have grown
throughout the world, the basic objective of pooling resources and
capabilities has really been reached.

Not everybody agrees with this statement as the US influence is still preponderant over any other view. This gives economic, political and moral problems and a truly international organization must provide strong guarantees to the smaller countries. Similar organizations which are desirable also for other applications face the same problems.

At this point it should be noted that Intelsat has no control over information, only the technical responsibility. The control is through the earth terminal transmitter and receiver making the system self-policing. But technology is moving quickly and as tomorrow's simpler and less expensive installations arrive, the new systems must be organized.

Here is perhaps the most pressing international problem. Transmission from satellite to home receivers is technically feasible today and economically realistic within the next decade. But there seems to be a lack of ability to grasp the situation and little progress is seen towards a solution. And as technology is proceeding fast, we may reach the future before we are ready.

THE IMPACT OF EARTH RESOURCES EXPLORATION FROM SPACE

William Nordberg

National Aeronautics & Space Administration

Goddard Space Flight Center, Greenbelt, Maryland, U.S.A.

1. INTRODUCTION

In contrast to the use of satellites for weather forecasting and for intercontinental communications, which have been available for more than fifteen years, surveys of earth resources with satellites have begun only a few years ago. The impact of earth resources surveys from space on society is therefore still much more a matter of speculation than a matter of record. Nevertheless, during three years of successful operation of the LANDSAT system, we have gained enough experience to assess the potential impact of global surveys with satellites of crops, forests, grasslands and soils on agricultural predictions: of snow cover and watershed morphology on water resources management; of land use, water and air quality on environmental protection: and of land forms and tectonic structure on mineral exploitation.

It appears paradoxical that one would want to go to outer space, to satellites orbiting many hundreds of kilometers above the earth, to learn something useful about our immediate environment. However, the ever increasing demands that are posed by the increasing population on the resources and environmental conditions of this planet, are making it mandatory to devise methods for the global assessment of the present as well as future state of our food, fiber, water, mineral and energy resources. Such assessments will be necessary to manage the extraction, distribution, conservation and renewal of these resources. Yet, it is evident that it would be prohibitively expensive to make such assessments frequently and over large areas of the world with conventional monitoring methods.

On the other hand, satellites can observe almost the entire
globe within a relatively short time period (12 hours to 18 days,
depending on the desired spatial resolution); also, satellite observations recur frequently and regularly for periods of many years
at very low incremental cost (the LANDSAT-1 system is still func-
tioning after three years of operation): and satellite observations
are made everywhere by the same instruments under the same controlled
conditions, so that versatile, thematic maps can be readily produced.

Satellites are, therefore, the only economical and practical
tool from which we may expect to obtain the kind and amount of information necessary for global management of earth resources and
for protection of our life-sustaining environment.

2.   THE   LANDSAT   SYSTEM

The one satellite system that has been used widely for the
purpose of surveying earth resources and of observing environmental
impacts, though only experimentally, is the LANDSAT system of the
U.S.A.   Two LANDSATs are now in such orbits that polychromatic
images in four spectral bands of reflected solar radiation can be
taken along one-hundred-and-eighty-five kilometer wide strips, anywhere in the world, every nine days.   But, LANDSAT-1, which was
three years old in July 1975, yields data only within broadcast
range of a receiving station; it cannot store pictures on its tape
recorders anymore.   Therefore, observations every nine days using
both LANDSAT's, are possible only where such receiving stations
exist.   This is the case for North America, via three U.S. and
one Canadian station, and over South America, via a Brazilian station: it will be the case soon over Europe and North Africa, via
the station that is just now being completed over Italy: and over
southwest Asia, southeast Europe and northeast Africa via a station
that is being procured by Iran.   A station proposed to be located
in Zaire, would cover most of the remainder of Africa.   Interest
in such receiving stations has also been expressed by a variety of
additional countries, so that the thematic mapping which can be
accomplished every nine days by the two LANDSATs is expected to
cover increasingly large areas of the world.   Other areas, which
are not within range of such receiving stations, must depend on the
data stored on the tape recorders that still function on LANDSAT-2.
Such coverage occurs, of course, much less frequently.   Namely,
one to several times per year, depending on interest expressed by
people in these areas.

The third LANDSAT, LANDSAT-C, is expected to be launched in
1977.   In addition to images in four bands of reflected light, it
will make images of surface temperatures at a somewhat lesser spatial resolution, namely, at about two-hundred meters.   Adding this

temperature mapping cability to the LANDSAT system, will improve
significantly our capability of classifying crops, vegetation and
soils.   Also for 1977 and in addition to LANDSAT-C, a heat capacity mapping mission (HCMM) is planned to map surface temperatures
from a satellite that will sweep the earth at hours of maximum and
minimum heating.   It is expected that soil moisture patterns, and
soil or rock compositions can be distinguished better with such a
satellite than with LANDSATs which do not scan the earth at times
that are optimum for this purpose.

Finally, we may expect that by 1980 earth resources surveys
will be improved with the development of a new thematic mapper.
This instrument would perform measurements with greater radiometric
accuracy, at six spectral bands, more appropriately placed, and
with greater spatial resolutions than the present LANDSATs.   However, a satellite mission to fly this thematic mapper has not yet
been approved in the United States.

Observations resulting from the LANDSATs have been analyzed
for a great variety of purposes by many hundreds of invidual investigators in more than forty countries.   These investigations
had been formally negotiated with NASA and many of their results
have been published.   Also, many governmental organizations,
especially in the U.S.A., Canada and Brazil have applied LANDSAT
observations as aids to their operations, such as sea and lake ice
surveys, preparations of environmental impact statements, water
runoff predictions, flood potential assessments and updating of
navigational charts, as well as of geological and general purpose
maps, especially in sparsely populated or rapidly changing areas.
In addition, several tens of thousands of individuals and organizations from all over the world have purchased LANDSAT data for
their private purposes from one of the public data centers which
the U.S. has established.   All this has resulted in an excellent
basis for assessing the potential impacts of earth resources surveys from space.

3.   POTENTIAL   IMPACTS

Information extracted experimentally from two LANDSATs, several
NIMBUS and the SKYLAB space flight missions has been used to demonstrate that a number of major problems which beset mankind today
could be addressed by surveys with satellite systems.   These problems are:

  1) Management of food, water and fiber resources.
  2) Exploration and management of energy and mineral resources.
  3) Protection of our life-sustaining environment.
  4) Protection of life and property.
  5) Improvements in shipping and navigation.

3.1 Management of Food, Water and Fiber Resources.   One of the
world's major concerns today is the production and distribution of
food crops.   World population projections indicate a need for a
three-fold increase in world food supplies to be distributed by the
year 2000.   World grain reserves have shrunk from twenty-six per
cent of annual consumption in 1959 to seven per cent in 1974.
North America has become the only major grain exporting region in
the world.

The management of the production and distribution of food could
be aided immediately by a global census and prediction of crop yield
and by a survey of rangeland conditions in major cattle feeding areas
of the world.   Such a census would be updated at frequent intervals.
LANDSAT observations have demonstrated that, given appropriate sampling from surface-based observations, most major crops can be identified in sequential satellite images and that acreage measurements
for these crops can be made with an accuracy of better than ninety
per cent.   Thus, a satellite based, global crop survey should be
feasible now.   Similar findings have also been obtained from LAND-
SAT observations of rangeland conditions.   The feeding capacity of
major grazing areas could also be estimated from satellite observations, if properly augmented with ground based information.   These
estimates could be used for the proper distribution and control of
livestock, particularly in grazing areas where marginal conditions
occur frequently, such as in the American West.

While results from past LANDSAT flights have proven the feasibility of making crop and forage inventories with satellites, they
have not yet resulted in any large-scale determinations of agricultural yield.   Investigations are now in process to make such yield
determinations, but they require a combination of observations from
various satellite systems, including weather satellites.   Such a
yield prediction system would combine the crop identifications and
acreage mensurations that are already being made with LANDSAT, with:

  1)   analyses of soil types which can also be obtained from
       LANDSATs;
  2)   information on cloudiness, insolation and rainfall which
       would be derived from existing meteorological satellites;
  3)   with soil moisture information that could be derived from
       passive microwave measurements such as are now being per-
       formed with NIMBUS or from radar observations that are
       expected to be available from early flights of the SPACELAB.

Information resulting from such satellite-based crop and forage
inventories and from yield predictions would result in both improvements in the planning for global distribution of grain crops and
improvements in crop and rangeland productivity.   The latter would
be achieved by improved decisions on planting, watering, fertiliza-
tion, pest control, harvesting, storage and on cattle feeding.

The former would be achieved through proper allocation of transportation requirements, earlier warning of crop failures, lesser
fluctuations in markets, more rapid fulfillment of requirements
and effective planning and monitoring of trade agreements.   Economic benefits from these plus from a stabilizing effect on the
commodity market have been estimated to range from about $10 million
per year to several hundreds of millions per year.

Another satellite-based observation capability which should
have a major impact on the management of food, as well as energy
resources, is the survey and prediction of availability of water.
The world's demand for water has increased markedly, as population
has increased and nations have increasingly industrialized.

In many areas of the world, such as the western United States,
the annual water runoff from snow melt is being used to fill these
demands.   Sequential snow cover observations from LANDSAT have
supplied information on the rate of melting and on the volume of
water released subsequently.   In ungauged, inaccessible and remote
watersheds, LANDSAT snowcover data which have been related to
seasonal runoff are the only means to provide early estimates of
water supply.   Even in areas where snowpack is monitored by conventional networks, satellite snowcover data have been used to
correct runoff forecasts.   These corrections often amount to
twenty-five per cent of total predicted runoff.

Since water runoff is managed by controlling the flow from a
series of reservoirs with flood prevention as a governing criterion,
better estimates of runoff accuracies will avoid the dumping of
water based on erroneous runoff predictions.   More accurate runoff
predictions, therefore, will result in more water being available
for irrigation, power generation and industrial use.   The value of
this improved efficiency in water utilization has been estimated to
be between twenty and fifty million dollars per year in the U.S.A.
alone.

LANDSAT observations have also been effective in recognizing,
locating and mapping the broadest classes of forest land and timber.
Deciduous, evergreen & mixed forest communities have been identified,
total acreages of various types of forest have been measured and
changes, such as produced by cutting or new seeding, have been
readily detected.   Under certain conditions, timber volume, age of
forests and presence of disease have been determined.   But, such
satellite surveys must be supplemented by on site and aircraft
measurements in judiciously selected areas to delineate changes and
patterns on scales that are not detectable by satellites and to
satisfy the needs of local forest management.   Such an integrated
monitoring system will have appreciable impact on regional logging,
thinning, reforesting and pest prevention operations, in addition to
national economic planning and conservation actions.   Estimates of

annual economic benefits from a combined satellite/aircraft/ground
sampling system for the U.S.A. range from $1 to $3 million for a
national forestry inventory to $27 million for a system that could
be applied to regional forest management decisions.

3.2 Exploration and Management of Mineral and Energy Resources.
The world's consumption of important minerals has increased more
rapidly than the population.   Known reserves of many critical
minerals range only up to twenty-five years and no new significant
and proven major deposits have been found since 1950 for many of
these materials.   Energy consumption has increased even more
dramatically and estimates of fossil fuel lifetimes to support this
growth, range from decades up to one-hundred years.   Exploration
for minerals and petroleum can be aided by satellite observing
systems which are capable of mapping synoptically the geomorphology
and general geological environment of very large areas.   Such mapping can also facilitate the siting of power generating plants thus
minimizing the cost of environmental protection and resulting in
lower cost for energy.

Analyses of geological lineaments observed by LANDSAT have
demonstrated that tectonic features such as faults or fractures can
be identified and mapped, and that these features may be indicative
of unexplored mineral deposits.   Landforms such as salt domes have
been identified and their relationship to petroleum deposits has
been established.   Colour tone differentiations observed by LANDSAT
have permitted the delineation of rock types, vegetation or soil
differences, each again leading to possible conclusions regarding
mineral deposits;   some of these colour tones have been related to
oxidation or other chemical reactions on the surface which may imply
mineralization or petroleum deposits.   A study of LANDSAT images of
northern Alaska has revealed an interesting alignment of lakes over
distances of several hundreds of kilometers which has led to specu-
lations about the extension of petroleum exploration in northern
Alaska.

The major impact of satellite observing systems on the mineral
and energy resources problem is a reduction in the time it takes to
complete the exploration cycle, lowering exploration costs by
reducing the amount of more expensive geophysical, geochemical and
seismic surveys;   and reducing the costs of analyzing the potential
environmental impact of operations related to mining and energy
generation.

3.3 Protection of our Life Sustaining Environment.   General population growth accompanied by extensive economic and industrial
growth has resulted in significant pressures upon our finite land,
water and air resources.   As far as land resources are concerned,
large areas will be consumed by urban sprawl, and some have

estimated that by the year 2000 the "urban impact zone" will include
a third of the area of the U.S.A.   The siting of nuclear power
plants, coastal zone development, mineral extraction from near surface, and the need for the preservation of wildlife, wildland and
agricultural land all call for complex land management decisions.
These decisions will depend on vast amounts of timely information
that can be provided economically only by satellites.   For example,
land use planning has been severely hampered by the lack of suitable
maps showing the types of land use over large regions and their
changes with time.   Such maps are needed on scales ranging from
1 : 25,000 to 1 : 500,000 and are outdated or unavailable even for
many parts of the U.S.A. because of the prohibitive costs of making
frequent surveys with conventional methods.   The situation for the
rest of the world is even less satisfactory.   Vast areas of Africa,
Asia and South America are poorly, and often incorrectly, mapped.
Over the last twenty years a project sponsored by UNESCO has completed a series of land use maps at scales of 1 : 5,000,000 to
1 : 20,000,000.   But, these maps are insufficient in detail to
assist developers and managers in many of their decisions.   Images
produced with the LANDSAT system have resulted in the capability to
map land use, practically anywhere in the world, at least once a
year on a scale as large as 1 : 250,000.   A detailed land use map
of a three-state area of Massachusetts, Connecticut and Rhode Island
in the U.S.A. was prepared from LANDSAT images and eleven separate
land use categories were identified and mapped.   The time it took
to compile the information for this particular survey amounted to
two to three months, which is one-tenth of the time it would have
taken with conventional methods, such as aerial surveys.   The cost
of making such a survey by conventional methods would have been
about one order of magnitude greater than that of making the survey
with satellite data.

Environmental impact assessment and improved reclamation planning has also been demonstrated with data obtained from LANDSAT.
For example, destruction of vegetation by fumes and timber cutting
due to copper mining operations around a ninety square kilometer
area in Tennessee, U.S.A., was observed in a 1 : 250,000 scale re-
production of a LANDSAT image.   Various degrees of vegetation
denudation could be clearly mapped.   LANDSAT images have also
revealed the denuding of vegetation in Ontario, Canada, where pre-
vailing winds carrying sulphur dioxide fumes from a local sintering
plant have caused similar damage.   In western Maryland, and in Ohio,
U.S.A., the extent of coal strip mining and the progress of reclamation is being monitored.   Acreage figures are being compiled for
seven land use classes, ranging from stripped earth to fully
reclaimed land.   This monitoring technique provides more timely,
accurate and complete data than are available from conventional
surveys on reclamation of the mined areas.   Satellite observations,
therefore, can be used to delineate and measure the degree of vegetation damage from industrial operations in a most economical, rapid
and objective manner.

   In developed and industrialized nations there are numerous
national and local laws, regulations and land use practices which
could not be enforced without the type of land resources information that is provided by satellites.   Estimates of economic benefits in the U.S.A. range from $10 to $115 million per year based on
varying assumptions, but not taking into account any improvement in
the "quality of life".   However, the need for satellite land use
surveys is not confined to industrialized nations.   Developing
countries have as much or greater needs for surveys of soil characteristics, water availability and vegetation cover to plan for
orderly developments and management of their land resources.

   Similar considerations hold for the need to improve and con-
serve water quality and to provide timely, effective and economical
information to enforce such conservation.   The Federal Water
Quality Act of 1972 was established in the U.S.A. to regulate the
discharge of effluents into the bodies of water of that country.
Other laws and regulations govern ocean dumping to protect the
offshore marine environment and the coastal zones.   Surveys from
space provide the broad view that is required for regional water
quality management and enforcement of national or international
regulations.   Satellite techniques have been and are being developed
so that they may be of value to those charged with the enforcement
of water quality controls.   As yet, only a few of the most important
water quality parameters may be monitored from space, namely, surface
temperature, algeal blooms, turbidity and sediment distribution, and
possibly large-scale petroleum spills.   Techniques are now being
developed for the remote sensing of more specific quantities relat-
ing to water quality and for mapping these quantities with greater
spatial resolution than was possible with LANDSAT.

   Satellite monitoring of air pollution is as desirable as is
monitoring of land use and water quality.   However, there are not
yet any systematic surveys of air quality with satellites that could
aid in the enforcement of air quality regulations.   The transient
nature of air pollution requires frequent observations of which the
LANDSATs are not capable.   Nevertheless, LANDSATs have detected
significant aspects of this environmental problem.   A highly polluted, stagnant air mass extending over an area of about two-hundred
by one-hundred kilometers was observed over an exclusively rural
area of the eastern United States during August 1973.   This is an
area in which no industrial or other pollution sources are located.
Yet the satellite had shown that such pollution had drifted in from
the highly industrialized areas some five-hundred kilometers to the
northwest.   Similarly, smoke plumes emitted from steel mills in the
state of Indiana, U.S.A., have been traced in LANDSAT images two-
hundred kilometers across Lake Michigan as they formed condensation
nuclei over the lake and caused heavy snowfalls on the other shore
in the state of Michigan.   Experiments to demonstrate primarily the
feasibility of monitoring air pollution from space on a more systematic basis will be conducted with NIMBUS-G in 1978.

3.4 Protection of Life and Property.   Aside from the property and
lives that can be saved through the use of satellites in achieving
more accurate forecasts of severe weather, there is a potential impact of space-based observations of geological and tectonic features
on the prediction of earthquakes, volcanic eruptions and on the planning of construction and reclamation projects.

     Preliminary experiments conducted with high precision laser
tracking of satellites and radio-interferometric tracking of quasars
have resulted in indications that such techniques might be used to
measure directly tectonic motions which are of the order of several
centimeters per year.   Such measurements could provide advance warnings of earthquakes much earlier than is now possible with conventional techniques.   However, it will take at least the next five to
ten years to refine our ability of making such high precision distance measurements between and within tectonic plates.

     On the other hand, mapping of regional lineaments observed in
LANDSAT images has already resulted in delineating seismic hazards
not otherwise recognised.   Investigators have discovered evidence
of recent movement along faults in an area of California that had
long been considered to be inactive.   A mosaic of LANDSAT images
of central Alaska has clearly revealed the presence of new sets of
lineaments that correlate well with the distribution of shallow-
focus earthquake epicenters in that region.   At least one volcanic
eruption has been observed by LANDSAT and the amount of debris dis-
charged into the atmosphere could be estimated from the size of the
plume which could be measured precisely in the image.   However,
LANDSAT observations occur much too infrequently to be useful for
the detection, let alone prediction, of volcanic eruptions.   Such
operations are depending on the LANDSAT data collection system
which relays seismic and other geophysical information from remote,
automatic and *in situ* sensors.

     LANDSAT images have been used to delineate geological hazards
to civil construction projects and to facilitate the planning of
such projects.   For example, an active leg of the Denali fault
was found to lie close to a proposed bridge site over the Yukon
River, and the proposed path of the Alaskan oil pipeline.   Sequential LANDSAT images of the Great Kavir area in Iran were used to
delineate seasonally wet, rough and unstable ground.   This analysis
resulted in the planning for a new road alignment that would considerably shorten the present route between northern and central
Iran.

     LANDSAT-1 images have also proven to be effective in identify-
ing surging glaciers and monitoring the areal extent of their change.
Surging glaciers can advance very rapidly over large land areas and
can cause devastating flooding by blocking and then suddenly releas-
ing large quantities of meltwaters.

3.5 Improvements in Shipping and Navigation.   The growing requirements for oceanic bulk transportation and the increasing size of
fleets as well as of individual ships (e.g. supertankers) will ac-
centuate the need for monitoring and predicting those factors which
affect the efficiency and safety of sea transportation; namely, sea
state, sea and lake ice distribution, ocean currents and hazards to
navigation.   The cost of a world-wide system utilizing buoys or
ships to measure these parameters on a fine enough grid scale would
be prohibitive.   Some of these parameters have already been observed
by remote sensing from satellites, sea and lake ice has been surveyed
by LANDSATs and the results have been applied to aid geophysical and
geological exploration in the Canadian Arctic and Great Lakes shipping in North America.   Bathymetry has been performed on LANDSAT
images and shoals, reefs and sedimentation patterns have been mapped
to update navigational maps, particularly in coastal zones.   In
fact, LANDSAT observations are now being used in the U.S.A. to update all navigational charts including those for air navigation.
Although ocean currents have been recognized and mapped with both
temperature observations from meteorological satellites and with
LANDSAT images, satellites have not yet produced the type of current
measurements that would be useful to shipping forecasts.   A similar
situation holds for satellite observations of sea state, the feasibility of which has been demonstrated in many aircraft flights and
by SKYLAB.   But, in contrast to sea ice surveys and observations
of navigational hazards, sea state and ocean current measurements
from satellites are not expected to have any major impact on maritime operations before the early 1980's.   Eventually, improvements
of ship routing based on such satellite surveys is estimated to
result in benefits of $30 - $50 million per year to U.S. trade.
Benefits to Canadian Arctic operations and offshore oil production
would potentially increase this amount by a factor of four.   Additional benefits would accrue from iceberg reconnaissance and the
optimization of ship routing from the Alaskan north slope oilfields.
Such surveys will also have a substantial impact on increasing
coastal activities, such as the construction of superports and on the
prediction of the potential environmental hazards of very large
supertankers.

4.   CONCLUSIONS

     Observations of phenomena and processes occurring on the earth
from satellite systems such as LANDSAT and NIMBUS have demonstrated
the potential impact of such observations on a number of major
human concerns.   These concerns include the management of our food,
water and fiber resources; the exploration and management of mineral
and energy resources; the protection of our life sustaining environment; the protection of life and property; and improvements in
shipping and navigation.   Satellite systems which are now under

development such as LANDSAT-C, SEASAT, NIMBUS-G and heat capacity mapping will have the potential of impacting those concerns even further and more specifically.   However, these impacts will remain potential rather than real as long as there is not an effective transfer of information from those who make the observations to those who are in need and in a position to apply this information to the solution of these problems.   I consider the achievement of such information transfer as a major challenge to the space programme of the next decade.

THE IMPACT OF EARTH RESOURCES EXPLORATION FROM SPACE

E. Tandberg, Norway

An edited summary of the discussion

SOME HIGHLIGHTS OF EARTH RESOURCES EXPLORATION FROM SPACE

To be able to judge the potential values of earth resources exploration from space, it is necessary to know how the space observations are made. Thus, for the benefit of the readers who may not have kept up with the science of remote sensing from space, we will give a very short introduction to the basic operational characteristics involved in the LANDSAT* system:

The satellite observations are polychromatic or multispectral to record different processes and phenomena that manifest themselves in different parts of the electromagnetic spectrum. Through their typical radiation, or signature, the individual processes and phenomena can be identified and analyzed much better than from visual observations or regular photographs. In other words, it is a question of recognition, analysis and mapping of specific features on the surface of the earth on the basis of their spectral characteristics. The LANDSAT system uses observations in four specific, relatively narrow bands of reflected solar radiation:

---

*LANDSAT, previously ERTS (from Earth Resources Technology Satellite), satellites launched by NASA July 23, 1972 (LANDSAT 1) and January 22, 1975 (LANDSAT 2) in an effort to develop instruments and procedures for an earth resources observation system. The sun-synchronous satellite orbits are circular, about 915 kms high with an inclination of 81° and a period of 103 min. Of the two principal earth resources observation instruments aboard the satellites – a return beam vidicon camera system and a multispectral scanner subsystem – only the latter will be considered in this chapter.

1)   in the wavelength of blue-green light (0.5-0.6 micrometers),
2)   in the wavelength of red light (0.6-0.7 micrometers),
3)   in the near infra-red region (0.7-0.8 micrometers), and
4)   a little further out in the near infra-red (0.8-1.1 micro-
     meters).

     Other satellites and other techniques go as far out as the
microwave-area to recognize certain features such as the distribution and the type of sea ice. Sea state signatures are in the
centimeter range.

     LANDSAT images can be displayed and more or less automatically
analyzed in each of the four bands of reflected solar radiation.
However, the images can also be combined into a simple colour com-
posite where a specific colour is assigned to each of the radiation
bands. It is possible to feed digital information for each of the
spectral images into a computer programmed to recognize statistical
characteristics in these spectral variations, and then assign the
characteristics to various features such as the water content of
snow, land use patterns, type of vegetation, etc.

     The ability of the satellite to view large areas is unique,
and there is no other way to recognize large scale tectonic features,
fault lines that extend over distances of two-thousand kms, etc.
To illustrate the usefulness of a broad perspective, consider for a
minute the Canadian and Alaskan tundra, where glacial processes have
left an immense amount of lakes of dimensions ranging from a few
hundred meters to some tens of kilometers. It has been known for
many years by just looking at individual lakes or groups of lakes
from airplanes that their shape is elongated. Each lake is elliptical, and its long axis follows the flow of the prehistoric
glaciers. Satellite pictures have brought out, however, that large
groups of lakes in areas stretching out two to three-hundred kms
are not oriented north to south as the glaciers were flowing, but
east to west. Geologists have indicated that this phenomenon is a
strong indication of the tectonic structure and can give valuable
information on the bedrock underneath the surface. As a consequence
it has been suggested that certain oil fields found in Central Alaska
extend along the alignment of the groups of lakes into an area of the
northern slope of Alaska where no oil has been discovered yet.

     A final point is the power of making repetitive observations
with a satellite. Whether it is every eighteen days as one LANDSAT
does or every nine days as two LANDSATs do, the regular, repetitive
and recurring nature of the observations, made with the same basic
techniques, do not require an intercalibration of the sensors and
methods that are being used in different parts of the world or even
in different parts of one country in order to arrive at a common
picture. The same sensors, the same systems make the measurements

everywhere and all the time in the same sun angle and for long
periods of time. One good example of using these repetitive observations is the prediction of runoff water from snow melt.

1. ACTIVITIES OUTSIDE THE UNITED STATES

As of September 1975 the Soviet Union has not launched any
special earth resources satellites.   However, earth resources observation experiments have been performed on unmanned spacecraft in
the Meteor and Cosmos series, on manned Soyuz spacecraft and on
Salyut space stations.   Black/white and polychromatic photography
in different spectral bands has been used from space, while multispectral scanning systems, infra-red scanners, microwave instruments
and different types of radars have been tested mainly from airplanes.
Some problems were quoted in the areas of data collection, data
transfer to users and computer analysis/identification.   Another
problem area was teaching the users the new methods, and to get the
results back.   Enough background and experience exist in the Soviet
Union to take on the next step - organizing a space/earth research
service.

In Sweden a remote sensing committee was created in 1969.
This committee has devoted considerable time and effort in telling
governmental bodies and scientists what remote sensing is and how
it can be successfully used.   The funds are limited, and the practical work has to be concentrated on a few promising areas such as:

a) Early detection of oil spills in the sea, also under bad
    weather conditions and in darkness.   This system is close
    to becoming operational.

b) Sea ice surveillance, also under bad conditions.

c) Registration of vegetation for use in agriculture, forestry,
    area planning, etc.

d) Pollution of the atmosphere.

The work is based on existing and well-established technology
like infra-red scanners and multispectral scanners, but the platforms
normally used are aircraft and helicopters.   The earth resources
satellites have generated a great interest.   On the other hand,
Sweden is a very diversified and rather well-studied country, so
that ground resolution of the LANDSAT images is normally not high
enough for new information in fields like geology, vegetation, mapping, etc.   There is also a need for better spectral resolution and,
since the weather in Sweden is often cloudy, especially in winter-
time, all-weather capability.

The European Space Agency has been looking at the activities in

the ten member states in order to find out if a pattern is emerging.
One of the conclusions is that priority must be given to the establishment of ground stations capable of receiving data – at the moment
principally LANDSAT data.   Attention must also be directed to the
education of possible users, getting them acquainted with the interpretation and utilization of data.   In the field of subsystems and
technology development the European Space Agency needs to concentrate
on an all-weather capability, which is possibly an area where there
is room for European initiative.   The Agency hopes that all this
will lead to the development of a satellite programme.   Those who
already have their earth resources observation programmes under way
are not likely to be the first to volunteer to give up money for a
new one.   Nevertheless it is hard to think that the number of
potential users in Europe could not justify a satellite in the
medium term future, and this is the aim that the agency will be
putting forward.   It will not be a programme looking around for
users.   The programme will be developed by the users in the process
of educating them and explaining what the possibilities are.

2. AIRPLANES OR SATELLITES?

The question of whether satellites were really needed for earth
resources observations when airplanes are available was raised
several times in the discussion period.

While it is true that airplanes are relatively economical (even
though a flight cost of about $1,000 per hour was quoted for air-
planes capable of flying at high altitudes), flexible, and relatively
simple, they offer limited coverage, have a limited range, and need
airfields.   There is no doubt that information on a local or even
regional scale can and should be gathered by an airplane or by send-
ing out surveyors, etc.   However, once a satellite is operational
it represents an almost infinite data source, and the greater use
the more economical the system will become.   This is the unique
characteristic of most large scale observations.   And there are
certainly tasks that the satellite can and the airplane cannot solve.
Some geological features can be detected or identified only by observations from satellite altitudes.   It is also very hard to com-
pete with the satellite when observing changing phenomena.   In
regional planning, and area becoming more and more important also
for developed parts of the world, the satellite is the only practical
source of observations for updating land use maps and statistical
records.   An airplane is often available when it is cloudy, and is
somewhere else when the weather is nice.   In most cases operational
satellite data cost one to two orders of magnitude less than data
collected with more conventional means such as airplanes.

3. GROUND STATIONS

In addition to the ground station projects and plans mentioned
by William Nordberg in his paper, the participants at the Symposium
were informed that France intends to help High Volta to transform a
telemetry station for earth resources satellite data.   The telemetry station is scheduled to close in 1975, and the conversion will
allow reception of LANDSAT observations in a part of Africa not
covered by the Italian station and the planned station in Zaire.
The High Volta station should be of help also because this part of
Africa has a considerable water shortage.

Norway is trying to raise funds to convert a telemetry station
at Tromsø into a LANDSAT receiving station, and is counting on support from the other Scandinavian countries.   The European Space
Agency will propose that a third European station is built in
Belgium.

An interest for ground stations has been expressed by countries
like Venezuela and Indonesia.

Fourteen ground stations would, if systematically placed, cover
almost all of the earth's land mass.

4. WILL SATELLITES IMPROVE HARVEST PREDICTIONS?

With LANDSAT 2, NASA, in co-operation with the U.S. Department
of Agriculture and the National Oceanic and Atmospheric Administration (NOAA) in the Department of Commerce, initiated the Large Area
Crop Inventory Experiment (LACIE) to test whether the use of data,
gathered by spacecraft and analyzed with the aid of computers, could
improve the timeliness and accuracy of major crop forecasts.   The
experiment combines crop acreage measurements obtained from LANDSAT
data with meteorological information from NOAA satellites and from
ground stations to relate weather conditions to yield assessment
and ultimately to production forecasts.   The Department of
Agriculture will study the experimentally derived production estimates for possible use in its crop reports, which are made public
to the domestic and international agricultural community.   At the
outset LACIE concentrated on wheat grown in North America.   If this
activity proves successful and useful, it will be extended to other
regions and ultimately to other crops.   Speaking before the U.N.
World Food Conference in Rome in the fall of 1974, Secretary of
State Henry A Kissinger called the experiment "a promising and
potentially vital contribution to rational planning of global production".   One of the eleven resolutions that was passed at the
conference called for the establishment of a global food monitoring
and assessment programme.   Not all countries agreed on that

resolution, but there was an overwhelming concensus among the
nations of the world that the programme was needed.

Several opinions were expressed at the Symposium.   One of the
participants pointed to a lack of real follow-up on the U.N. World
Food Conference resolution, which within the context of a world
approval, had enough room for the organization and planning of an
effort perhaps through the establishment of a single international
operating agency.   Others doubted the ability to forecast crop
yields at all, because of unknown weather conditions, pest factors,
etc.   It was also stated that the resolution, instead of improving
world agriculture and the food situation, would only increase the
number of "agricultural bureaucrats".

On the other hand William Nordberg hinted that LANDSAT data,
if processed properly, would improve wheat predictions by a few per
cent and a few months in time.   In other words - instead of predicting in May with an accuracy of ninety-five per cent what the
wheat harvest will be in the United States, based on the present
reporting system which is probably the most sophisticated in the
world, it could be possible to predict in March with ninety-six or
ninety-seven per cent.   This would be of considerable benefit.

The United States has an agreement with the Soviet Union to
exchange conventional data of harvest predictions.   This exchange
takes place, but by putting satellites to work the data could be
improved by a substantial increment.   Other wheat exporting countries, like Argentina, do not have very efficient reporting systems
at all.   To institute a high precision, long term reporting system
just for wheat would be very expensive if conventional methods are
used.   Satellites could be the solution, and when the technology
now being developed for wheat is used on other crops like rice,
soybeans, corn, etc., it could have influence on the global problem
in food production and food distribution.

5. NEW SENSORS

Part of the technical discussion was devoted to the area of
further sensor development.

Even though NASA is a space research and engineering oriented
and not an information transfer and user oriented organization,
William Nordberg held that the development of instruments and techniques with the engineering thereof was second priority to the information transfer.

The multispectral scanners of the LANDSAT system were praised,
but had a limitation in the accuracy of measuring the radiation
intensity.   Actually, the limitation is not so much in the sensor

itself as in the telemetry system.   The scanner permits the measurement of the radiances within a general accuracy of one per cent, but
with variations from one band to another.   The LANDSAT telemetry
system, on the other hand, permits transmission only within an
accuracy of about two per cent.   An analysis of aircraft remote
sensing information points to an accuracy of about 0.5 per cent for
categorizing crop and vegetation patterns, thus indicating the need
for an improvement in the LANDSAT system.

Another improvement should be made in the sensor ground resolution from the present approximately eighty meters - not to five to
ten meters because this would give tremendous data handling problems,
but to thirty to forty meters - for agricultural classifications.
There is also a need for narrowing and optimizing the spectral bands
and probably adding one more band, around 1.6 micrometers, again for
agricultural classification.   This is in essence the thematic
mapper which will be ready to fly on a satellite after LANDSAT-C if
the United States should decide to continue the LANDSAT programme.

Going beyond the polychromatic mapping of the LANDSATs, the
next important question is the reliable survey from space of soil
moisture on a broad scale.   To estimate the yield - the production
in tons of wheat, rice, grain, etc. - soil moisture is one of the
most important parameters.   Soil moisture information can be
gathered several ways: one is to simply study the heat capacity
of the ground, with measurements of temperature day and night in
the infra-red 11 micrometer window whenever it is cloudfree.   With
other necessary background information the temperature variations
will relate to soil moisture.   Another method would be to use
radars or passive microwave techniques to map the radio emissivity
of the surface, which is a function of moisture.   Such microwave
techniques for measuring soil moisture are actually being developed,
and are in the process of being tested on aircraft in both the
United States and the Soviet Union as part of a co-operative programme agreed on by the two nations.

Another interesting area would be measurements of water properties - water quality as well as the biological productivity
(chlorophyl content) of water.   At present this is beyond the
capability of the LANDSAT system, but there is a possibility that
shorter wavelengths, images in the blue part of the spectrum, can
provide such information.   These measurements are limited by a
scattering effect in the atmosphere, but there are solutions to
this problem, and a scanner is now under development in a NASA
laboratory at Goddard Space Flight Center to measure, in six or
seven bands, the shorter wavelength components of reflected sun-
light over water and possibly derive from it an estimate of
chlorophyl, particularly in the coastal zones.   The scanner will
be flown on the NIMBUS-G satellite.

Next would be measurement of air quality, the identification of pollutants in the atmosphere, first of all qualitatively, then, more importantly, quantitatively, particularly in the lower atmosphere. This is exceedingly difficult, but radiometric and spectrometric techniques in the optical part of the spectrum are now under development in the United States as well as in the United Kingdom (treated also in the next chapter).

Mentioned in the discussion were also all-weather sensors, or instruments with the ability to look at surface features largely insensitive to obscuration by clouds. A very coarse resolution microwave sensor is now operational on NIMBUS 5 and NIMBUS 6. On NIMBUS 5 it is used to map the world at wavelengths of 1.55 centimeters with a ground resolution of about thirty kms. The data give primarily precipitation patterns and ice patterns. On NIMBUS 6 the same mapping will be carried out at eight millimeter wavelength. Perhaps a combination of the various passive microwave techniques could also give information on soil moisture, but active techniques (radars) appear most promising for this purpose.

Some scientists are enthusiastic about the thought of mapping the earth's resources from a geostationary orbit, and propose to do so with a one to two meters telescope operating in the thermal spectral band plus the same bands as LANDSAT. The observations would have about the same ground resolutions, but would be continuous with a better potential than LANDSAT for detecting forest fires, floods, etc., and observing coastal zone processes, vegetation, wild life migration and the like.

6. ONE NATION OR AN INTERNATIONAL ORGANIZATION?

Several participants voiced concern because LANDSAT, the only known advanced earth resources satellite system is operated by the United States alone. In their opinion an international organization should control all satellite functions and the data handling to prevent possible misuse - information on natural resources belongs to the individual countries because of the economic and political aspects involved. A few participants feared that the United States might suddenly stop operating the system or change instruments in the satellites after considerable funds were invested in ground stations and ground station equipment. One participant also cited the long term negative effect of leaving advanced science and technology to the United States.

There is no doubt that the dependence on a single nation's satellite system is not an ideal situation. However, the participants at the Symposium were all grateful to the United States for making the LANDSAT system available. After all, the use of the

satellites up to now has been free, and the multispectral images
are available to anyone at cost price.   No international organization could have developed a system comparable to LANDSAT in as
short a time span as the United States.   And as far as the operational aspect is concerned, the view of several participants was
that the introduction of an international organization even at this
stage was premature because too little experience, too little confidence and too little evaluation of the potentiality and value of
the technique were found outside the United States.   The operation
of an earth resources satellite system is fundamentally different
from the communications satellite and the meteorological satellite
systems.   It would be impossible to have an international communications system without a prior agreement, because at least two
parties are involved.   With the earth resources and the meteorological satellite there is, in principle, only one.   However, since
the earth resources satellite has the capability of acquiring information of economic and political importance, there is also a
difference between this type of satellite and the meteorological
satellite.   In the world of meteorology there are long traditions
of data exchange and a completely free data distribution.   Most
people are interested in meteorological data, and it does not matter
whether it comes from satellites or more conventional sources.   The
conclusion is that it will be considerably more difficult to establish an international organization for operation of earth resources
satellites than communications or meteorological satellites.   And
even an international organization cannot guarantee that services
will not be interrupted at some future stage.   Nevertheless, international operation should be encouraged.

7. EARTH RESOURCES SATELLITES AND THE DEVELOPING COUNTRIES

    Brazil was cited as a good example of how earth resources
satellite data could be put to use in the third world.   This
country demonstrated an early interest through a strong participation in the LANDSAT-1 investigations, and established the first
ground station after Canada outside the United States.   It was
stated at the Symposium that in Brazil LANDSAT data are now being
used operationally for purposes like mapping the boundaries of the
country, geophysical exploration and environmental protection.   In
Bolivia special investigations resulted in a geological and geomor-
phological map unexistent until that time.   In Mexico LANDSAT data
have been used to map watersheds and vegetational parameters to an
extent never done before.   Similar investigations have been carried
out in countries from Africa to South-East Asia.   In Thailand, for
instance, work on resources inventories is being supported by international organizations such as the FAO or the World Bank.

    Generally speaking, there is in most developing countries a

real demand for maps of various kinds.   At least for maps with the
scale of 1 : 250,000 or smaller, this demand can be met by a systematic use of LANDSAT imagery.

However, what is needed in the third world are also good training programmes, training seminars or educational programmes.   These
training activities should stress how to get hold of information and
how the information can be used - for instance, to irrigate the
fields and to protect the pastureland.   Several international
organizations and development agencies in industrialized nations
have funds, and should be encouraged to play a more active role
than they have.   On the other hand, it may be necessary to educate
the responsible people in these organizations or agencies first.

8. CREDIBILITY AND LIMITATIONS

Most space projects are oversold, more or less intentionally,
at some stage of their development.   Application satellites are no
exception - in the early 'sixties there was almost no limit to what
the meteorological and communications satellites were imagined capable of doing.   The real impact of the meteorological satellites
were realized about three years after TIROS-1 was launched on
April 1, 1960, but it was not until 1966 that the Environmental
Science Services Administration in the United States started to
operate a meteorological satellite system for routine weather forecasting.   The first commercial communications satellite, INTELSAT-1,
or Early Bird, was launched on April 6, 1965, thus it is natural
that the earth resources satellites are characterized by some overselling in the time period before their real impacts are realized.

In an effort to find real users of LANDSAT data, the National
Oceanic and Atmospheric Administration was mentioned as an example.
The National Oceanic and Atmospheric Administration is, among other
things, responsible for all nautical charts, hydrographical surveying,
flood warnings and flood forecasts in the United States.   It is
using snow information from LANDSAT in run-off estimates, and is
using data for updating and correcting hydrographic charts.   Another
example is the United States Corps of Engineers in their effort to
survey all bodies of water of a certain size.   This survey would
have been extremely difficult to carry out without satellite data.
The Corps of Engineers is also involved in flood analysis based on
LANDSAT imagery.

Of course the LANDSAT system has its limitations.   For instance,
it is not capable of mapping air pollution.   The satellites were not
designed to do that, and do not have the right sensors to distinguish
between e.g. SO_2 or haze.   On the other hand, intensely polluted air
masses have been detected on LANDSAT pictures in several instances.
In general terms it can be stated that the limitations of the

satellite system is less of a sensor or a technological problem
than a problem of combining satellite observations with conventional
observations and with existing knowledge and models of what goes on
in the atmosphere and on the surface of the earth.

9. THE REAL VALUE OF THE EARTH RESOURCES SATELLITES

It is not economically feasible to develop a satellite to
solve only specific local problems - there are too many conflicting
requirements, and the data can be collected by simpler methods.
However, once the LANDSAT system was available, NASA found that the
data were actually more beneficial and more profitable - in numbers,
not dollar value - on the local scale than on the regional, national
or global scale.   For instance, there have been a number of cases
where LANDSAT data were used in the litigation of legal procedures
connected with discharge of pollutants.

Almost all thorough studies of mankind's future have come to
the conclusion that we will not succumb due to lack of resources.
The real difficulty facing the world as a whole is the management
of the resources.   The earth can probably support ten-billion
people if the resources are managed in a more sensible way.   This
is basically a political problem, but before it can be solved the
politicians need scientific background and data on a global scale.
The earth resources satellites could provide just that, and here is
the real, long term value of LANDSAT type systems.

In the words of Arthur C Clarke: "I think that what we have
here is a tool so enormous and so powerful that it's just going to
take us at least a decade to understand what we can do with it, and
it may take another decade for us to teach the world what can be
done with it.   And so I don't think we should be too discouraged
by the perhaps lack of appreciation at this stage.   I am sure that
in another generation we will not be able to imagine how we could
have run this planet without this kind of a satellite."

THE ENVIRONMENTAL SATELLITE: WHAT IT MEANS FOR MAN

Robert M. White

Administrator

National Oceanic and Atmospheric Administration, U.S.A.

1.   THE   USE   OF   THE   SATELLITES

The 1975 Nobel Symposium has posed for its theme a very large topic: the impact on mankind of space science and the applications satellite.   The satellite is one of the most glamorous triumphs of modern technology, and I venture to say that none of the participants in this Symposium will ever lose his sense of wonder that man can loft a large piece of hardware thousands, or tens of thousands, of kilometers into the atmosphere and place it in orbit around his own planet.   But however awesome the achievement, the satellite by itself is just that: a piece of hardware.   What is important is how it is used, and the final justification of the satellite is its ability to serve man in beneficial ways.

Over the past decade, satellites have been continuously refined and developed and extended to new uses.   Today we use the satellite routinely as a platform to relay communications signals between continents.   The satellite also provides a platform from which we can take an inventory of some of the surface resources of the entire globe, and perhaps obtain a revealing view of those earth formations that indicate what may lie under their surface.   Within the United Nations, governments are now discussing the international principles that should govern the use of satellites to observe the earth's resources and to broadcast television programmes directly to home receiving sets, although the latter application is still some years away.

Others will discuss these applications.   My own topic is the use of the satellite to monitor the physical environment, by which

I mean the atmosphere, the oceans and the solid earth.   What began
as a "meteorological satellite" has become an environmental monitoring satellite and for good reason.   The atmosphere, the oceans and
the solid earth form a single geophysical unit whose parts are constantly interacting.   The interactions between the oceans and the
atmosphere vitally affect the rate at which thermal energy, momentum
and water vapour are transferred to the atmosphere, and through
these physical processes determine the global weather.   In turn,
the currents of the oceans, which are generally massive and slow-
moving, are in large part a response over a long period of time to
the atmosphere.   And as the atmosphere moves over the solid earth,
it is continuously being heated or cooled by the surface below.
And so if we wish to understand the physical processes that shape
the weather and to keep a ready watch on the weather as it evolves,
we must monitor the physical environment in a unitary way.

We also want to monitor the physical environment for more than
the weather.   For ocean forecasting, we want to know about the
state of the seas, the flow of ocean currents and the formation and
break-up of ice in the more northerly and southerly latitudes.   We
want to know about snow packs, particularly for the management of
water resources and the forecasting of river floods.   And we need
to monitor solar phenomena so that we may predict propagation conditions for radio transmission and radiation conditions for manned
exploration of outer space.   I shall add to this list in the course
of this paper, but these few illustrations make it clear why the
meteorological satellite has become the "environmental satellite".

The satellite has a unique role to play in the work of the
environmental sciences.   It makes it possible – for the first time
– for man to observe his entire global environment.   The oceans
cover approximately seventy per cent of the surface of the globe;
and before the development of the satellite, observation of the
atmosphere over the oceans and of sea surface conditions depended
primarily on ships at sea.   At best, the process was sporadic and
confined primarily to shipping lanes;   and the data gathered were,
to say the least, sparse.   Another ten per cent of the earth's
surface comprises mountains, jungles, deserts and polar regions,
and it was equally inaccessible.   Here at best there was an
occasional observing station.   The environmental satellite has
changed all this – and dramatically.   The satellite has made the
entire planet accessible to observation routinely and with regularity.
Our ability to observe the environment is far from perfect;   the
equipment that we place on the satellite to do the observing is still
in the process of development.   But the important point is that the
satellite is an environmental observing platform that can scan the
complete surface of the globe and do so on a daily or more frequent
basis.

The satellite also makes it possible to measure the fields of

environmental parameters in a new way.   The traditional method has
been to measure the environment at various points and to deduce the
field from a series of point measurements.   With the satellite, the
environmental scientist can obtain an integrated view on a global
basis from the vantage of space.   We are still not satisfied with
our measuring instruments and techniques; but the capability is
there, and it is uniquely due to the satellite.

     In the pages that follow, I shall try to trace the history of
the environmental satellite and its principal on-board equipment,
assess the environmental satellite's contribution to mankind, and
discuss some of the questions about the control and management of
environmental satellites that are now being raised within the international community.   But let me anticipate my conclusion.   In my
view, the environmental satellite has revolutionized man's ability
to monitor the physical environment.   The full effects of this revolution have not yet been felt; in fact, they will not be felt for
many years.   But the satellite has already contributed materially
to the advancement of our basic knowledge of the properties and processes of the physical environment.   It has already proven invaluable in the protection of life and property against severe storms
and other hazards of the environment.   And it has already enabled
a large number of nations to improve their regular weather services
- services that are important to many of man's economic activities
and social pursuits and that intimately affect the quality of life
on our planet.

2. HISTORY OF THE ENVIRONMENTAL SATELLITE

2.1 TIROS.   The United States launched its first environmental
satellite in 1960.   It was experimental and was called the
Television and Infra-red Observation Satellite, or TIROS.   It was
launched into an inclined orbit that permitted it to view the broad
belt of the globe between the 50° parallels.   TIROS-1 carried two
television cameras, which took video pictures of whatever surface
was immediately visible.   Where clouds had formed, the pictures
were of the tops of clouds; in cloud-free areas, of the earth's
surface.   The pictures were stored in the satellite and transmitted
to two special receiving stations in the United States as it passed
overhead.   These receiving stations saw the satellite for only a
short period of time, and the video pictures the satellite had taken
and stored in its global orbit had to be transmitted to them at very
high speed.

     The video pictures taken by TIROS-1 literally revealed the
"weather".   It was as if a textbook on storms had come alive.
They showed the meteorologist where storms were in progress or in
development, and they indicated frontal activity and other significant phenomena.   And they gave him a total view of storms through

their cloud forms, not a fragmentary one.   The pictures were almost
startling in their clarity, and the weather services of the United
States recognized at once that here was a new technology that could
play a significant role in the task of describing and predicting
the weather.

Over the next five years, the United States experimented exten-
sively with the environmental satellite.   In this period, eleven
experimental satellites were launched – ten in the TIROS series and
one NIMBUS satellite.   One of the first tasks was to integrate the
video pictures taken by the satellite into the everyday process of
weather forecasting.   Meteorologists quickly learned how to use
the new satellite data in the preparation of their forecast charts,
and the TIROS pictures began to be used operationally, although only
in a limited way.   The pictures were particularly useful in providing a watch on the formation and movements of severe storms, notably
the late-summer hurricanes that are spawned over the Atlantic Ocean
and the Caribbean Sea.

The experimental programme concentrated considerable effort on
the improvement of the TIROS satellite itself.   On TIROS-1, the
video cameras pointed at the earth's surface for only about twenty-
five per cent of the flight;   during the remaining portion of the
orbit, they pointed uselessly at outer space.   TIROS-IX remedied
this defect.   It was designed to "roll" along its orbital path (in
effect, on a wheel), and with each spin the cameras pointed at the
earth's surface.   The upshot was greater picture-taking capability
– four times that of TIROS-1.

There was experimentation with the on-board equipment.   Both
narrow-angle and wide-angle cameras were used on various satellites,
and over time it became clear that the wide-angle camera was the
most useful for meteorological purposes.   The United States also
began to experiment with infra-red radiometers.   Video cameras
were usable only when the earth's surface was lit by the sun.   On
the dark side of the earth during each orbit, they showed nothing.
If there was to be complete global coverage in each orbit, there
had to be a night-time picture-taking capability, and the infra-red
radiometer could provide it.   It measured the temperature of the
surface that was immediately scanned.   Since the temperature of
clouds varies with their height, the thermal pictures taken by an
infra-red radiometer could then be translated into pictures of cloud
cover.   An infra-red radiometer was also developed to measure the
solar and thermal radiation from the earth's surface and atmosphere
in order to provide data on the heat balance of the planet.

Another significant development of this experimental programme
was automatic picture transmission (APT).   There has always been a
tradition of co-operation among nations in the exchange of weather

information, and from the beginning of its environmental satellite
programme the United States has stood willing to make the data
collected by its satellites available to other nations.   But for
many nations, notably those in the developing world, the offer was
meaningless.   The cost of communicating satellite data from the
United States to other countries made transmission prohibitive, or
nations lacked the technical capability to process and analyze the
data.   APT is a partial solution.   A video picture is taken and
then slowly scanned for immediate transmission to earth.   The
transmission is made on a relatively narrow bandwidth so that the
signal can be "read out" by a relatively low-cost ($20-50,000)
receiving set.   The APT channel of a satellite is in continuous
operation; it is never shut off.   APT thus permits any nation, at
little cost to itself, to secure an instantaneous picture of its
overlying atmosphere as a satellite equipped with APT crosses during
its orbit.

     TIROS-1 was conceived as a meteorological satellite in the
narrow sense of the term.   Its video cameras were designed to scan
the weather - to look at what was happening in the atmosphere.   But
it was quickly obvious that the same cameras might tell us something
about the oceans and the solid earth, and the United States began to
experiment with the detection of sea ice and snow cover.   Clouds
are transient; snow and ice change slowly.   It soon became possible
to distinguish the white of snow and ice from the white of clouds.
The meteorological satellite was becoming an environmental satellite.

     Early in 1966, the United States launched a satellite system to
serve meteorology regularly and routinely.   While individual spacecraft have come and gone, and their on-board equipment has been
improved and supplemented, the system has operated without inter-
ruption since then.   This first operational system comprised two
TIROS Operational Satellites (TOS) and embodied the technology that
was then proven.   The satellites were launched into a polar orbit
and passed over the entire globe at least twice a day - about every
two hours over the poles, with decreasing frequency as their orbital
paths brought them over the equator.   One satellite carried two
wide-angle video cameras capable of taking four-hundred-and-fifty to
five-hundred pictures a day over the sunlit portions of the earth.
It also carried a low-resolution radiometer to gather data on the
heat balance of the earth-atmosphere system.   The satellite stored
the pictures and data it gathered as it moved along its orbital path
and then transmitted them to two large receiving stations - one in
Alaska and the other at Wallops Station off the coast of Virginia.
From there, the pictures and data were retransmitted to Washington
for processing and analysis.   The second TOS carried the cameras
for the APT system.   These operational satellites were called
Environmental Survey Satellites (ESSA).

2.2 ESSA.   The ESSA satellites provided an operational system of considerable value, but it was still a system of limited capabilities.   Modern weather forecasting is done numerically, with the aid of high-speed computers; and the most significant parameters for weather prediction are the vertical temperature structure of the atmosphere, the motions of the winds (both their speed and direction), and the moisture content of the atmosphere.   The ESSA system provided no direct data about these factors.   The United States was well aware of this limitation; and during the latter half of the 1960's, it turned to the development of new instruments and techniques to overcome it, using the NIMBUS satellites of the National Aeronautics and Space Administration.   It developed remote sensors that could be placed on a satellite to "sound" the atmosphere on its vertical temperature distribution and moisture content.   It explored new techniques to determine the speed and direction of the winds.   And it devoted considerable effort to the development of radiometers that could provide better pictures (that is, pictures of higher resolution and finer detail) than were being provided by the video cameras carried by the ESSA satellites - and do so both day and night.

2.3 NOAA.   In 1970, the United States began to phase into its polar-orbiting operational environmental satellite system a second generation of satellites.   This was the Improved TIROS Operational Satellite (ITOS) series.   In operation, these satellites have been designated the NOAA satellites.   The name is the acronym of the National Oceanic and Atmospheric Administration, the agency of the United States Government that operates the American environmental satellite system.

The NOAA series constituted a marked advance over the ESSA series.   The ESSA operational system had required a separate satellite for the APT function and for the function of observing the global atmosphere, storing the observational data, and later transmitting the data to the receiving stations in Alaska and Virginia. With the NOAA series, one satellite performs both functions.

A NOAA satellite also carries better equipment.   On NOAA-2, scanning radiometers replaced cameras for regular observational work.   This radiometer scans the underlying surface in both visible light and infra-red.   Thus for the first time the operational system could provide pictures of the earth's cloud cover both day and night.   In addition, a NOAA satellite carries a very high resolution radiometer, which also senses in both visible light and infrared and which provides pictures of finer detail than does the scanning radiometer.   The higher resolution (0.8 km) is particularly useful for non-meteorological applications.   For the oceanographer, this Very-High-Resolution Radiometer (VHRR) provides data about the temperature of the sea surface and the presence of sea ice.   For

the hydrologist, it maps mountain snow and water resources.   Since
only a few images can be stored for later transmission, all the pictures taken by the Very-High-Resolution Radiometer are immediately
broadcast for reception by ground receiving sets.   The service is
known as High-Resolution Picture Transmission (HRPT).   The HRPT
receiving set is larger and more costly ($250-300,000) than the APT
receiving set used with the scanning radiometer (and earlier, with
the video camera).

     The NOAA satellites carry two other important pieces of equip-
ment.   One is a vertical temperature profile radiometer.   It
sounds the atmosphere at various levels, and the soundings give us
a temperature profile.   This same radiometer also tells us about
the moisture content of the atmosphere, although here the measurement is less precise than in the case of temperature.   The other
instrument is a solar proton monitor to measure the arrival in the
Van Allen belt of these energetic particles from outer space.   The
measurements help us to predict radio disturbances and to monitor
the radiation environment for manned space flights.   The NOAA
satellites were the first to make regular operational use of these
two instruments.

2.4 Geostationary Satellites.   While it was deploying its polar-
orbiting environmental satellite system, the United States began to
explore the use of the geostationary satellite for the monitoring
of the physical environment.   The virtue of this satellite lies in
the choice of orbit.   It is placed in an orbital slot thirty-six-
thousand km above the equator, and at this height its flight is
synchronous with the rotation of the planet.   The result is that
it is relatively immobile in relation to the earth's surface and
can provide continuous coverage of the same portion of the globe.

     Continuous coverage is particularly important for the detection
and tracking of severe storms that are small in size and transient
in nature.   Tornadoes, which are among the most destructive weather
hazards in the United States, are an excellent example.   They
develop and strike quickly, with little time for warning - certainly
in less time than it takes a polar-orbiting satellite to complete a
single orbit.   Continuous coverage is also valuable for tropical
storms, such as the hurricanes that develop over the Atlantic and
Caribbean and sometimes move over the continental United States.
It is true that these storms have a relatively long life - one that
is measured in days, not hours.   But a polar-orbiting satellite
passes over any particular point in the tropics only once in every
twelve hours, and these tropical storms can alter significantly -
in the velocity of their winds, in their forward movement, and in
their direction - between the passes of a satellite.   A geostationary
satellite can keep them under constant surveillance.

The United States experimented with the use of a geostationary satellite for environmental monitoring in the late 1960's and early 1970's. In 1974, it launched its first geostationary satellite for operational use; and earlier this year, it launched a second. One is in orbit over South America; the other, over the eastern Pacific south of California. Together they cover North and South America and the adjacent ocean areas (but not the polar regions). These two satellites are called Synchronous Meteorological Satellites (SMS). Later ones in the series will be called Geostationary Operational Environmental Satellites (GOES).

2.5 GOES. The GOES satellites carry a new radiometer called the Visible and Infrared Spin Scan Radiometer (VISSR). It provides both visible and infra-red pictures, although the nighttime coverage has less resolution than the daytime. The radiometer scans a latitudinal swath in the visible channel approximately one km wide with every spin of the satellite, and it takes about twenty minutes for the radiometer to scan the full portion of the earth within its view. If a severe storm is in formation or is already raging and we want to follow it very closely, GOES can provide a picture of it every fifteen minutes – or even every five minutes. But for greater frequency of pictures, the satellite must trade a degree of territorial coverage. The GOES satellites also enable us – for the first time – to measure the velocity of the winds. We do this by tracing selected clouds imaged at intervals of an half-hour and deriving the direction and speed of their movement.

The GOES satellites also carry a new Space Environment Monitor (SEM), which measures the earth's magnetic field, energetic particles (protons) at the satellite's height above the earth, and the x-ray portion of the solar spectrum. The data help us to forecast radio propagation conditions (for telecommunications) and the space environment (for very high-flying aircraft and manned spacecraft).

Finally, the GOES satellites provide two specialized environmental communications services. One is to collect environmental data from surface observing platforms that are relatively inaccessible: ocean buoys, automatic weather stations in remote areas, rain gauges in the mountains and river valleys, tide gauges and seismometers. These platforms are equipped with a special radio transmitter and send their data up to the GOES satellite; the satellites, in turn, retransmit the data to a central receiving station at Wallops Station, Virginia. The second service is a broadcast service called WEFAX. Here the GOES satellites serve as relay stations for the broadcast of weather information of interest to the Western Hemisphere (or more precisely, the portion of the globe irradiated by the GOES transmitters). The information is in the form of satellite pictures and weather maps, and the broadcasts can be received by anybody who has the requisite receiving set. WEFAX

serves ships at sea as well as land stations, and for nations in
the Western Hemisphere it offers an important supplement to the
APT service provided by the NOAA satellites.

NOAA and GOES, the one polar-orbiting and the other geostationary, together constitute the total operational environmental
satellite system currently operated by the United States.   The GOES
series will continue to be the operational geostationary satellites
at least through the early 1980's.   But the polar-orbiting satellites will change.   In 1978, the United States will introduce into
its system a third generation of polar-orbiting satellites - the
prototype of which is called TIROS-N.

2.6 TIROS-N.   TIROS-N will carry an advanced radiometer that combines the functions of the present scanning radiometer and Very-
High-Resolution Radiometer and will have two additional spectral
channels.   In addition, TIROS-N will have improved sounding equipment.   The present vertical temperature profile radiometer operates
in infra-red, and infra-red cannot penetrate the clouds.   This is
the serious drawback of infra-red instruments.   TIROS-N will have
an improved sounding radiometer.   It will be better able to obtain
measurements in partly cloudy areas.   The same instrument will also
provide better measurement than we now obtain of the moisture content
of the atmosphere.   TIROS-N will also carry a microwave sensor.
Microwave sounding is one of the major developments of the current
decade.   Unlike infra-red, microwave can penetrate the clouds;   and
microwave sensors will eventually give us uninterrupted sounding of
the atmosphere for temperature and moisture content despite cloud
cover.   Microwave sensors will be particularly important for the
polar regions and for the area just north and south of the equator,
which are consistently cloudy.

Finally, TIROS-N will carry two important pieces of equipment
being developed by other nations.   One is a British instrument for
the sounding of the higher levels of the atmosphere.   The other is
a French instrument that will track and collect data from moving
platforms - such as drifing ocean buoys and constant-level balloons.

The TIROS-N series will be phased into the operational system.
But the work of research and development on both satellite technology and the technology of the satellite's imaging and sensing
equipment is an ever-continuing process.   From the NIMBUS-6 satellite, the United States is now experimenting with sensing equipment
designed to measure the earth's radiation.   The sensor measures
three different factors: incoming radiation from the sun (the
primary source of energy for the planet), outgoing solar radiation
reflected from the clouds and the upper atmosphere (the albedo), and
outgoing radiation reflected from the earth's surface in the infra-
red portion of the spectrum.   These measurements have many uses,

but they are crucial to understanding climatic changes.   Thus far
we have not been able to make the measurements with enough accuracy.
Still another important development will be that of sensors to
measure atmospheric pollution, particularly in the upper layers.

2.7 SEASAT.   Before the decade is out, the United States will
launch a new experimental satellite: SEASAT.   It will serve
several disciplines - oceanography, meteorology and geodesy.   It
will carry an altimeter that will give us the geoid: the shape the
oceans assume under the influence of gravity, winds and the rotation
of the earth.   The altimeter will tell us the relative variations
in the elevation of the ocean's surface to within less than half a
meter.   SEASAT will also carry a radar scatterometer, which will
provide information on the ocean roughness and thus on the speed
and direction of the winds immediately over the oceans.   We can
now measure the atmospheric winds from geostationary satellites by
tracing the movement of clouds, but we cannot yet measure the winds
at the sea surface.   Since departures from the mean geoid are
associated with ocean currents and wind stress, we can use these
measurements to learn about the currents themselves.   SEASAT will
also carry a microwave radiometer to measure the temperature of the
sea at its surface even in the presence of clouds.   In addition,
its altimeter will measure the height of significant waves.

     The vital work of research and development now stretches well
ahead.   It is the necessary work of bringing the satellite into
full flower as an environmental observing platform.

3.   WEATHER   FORECASTING

3.1 Over the past fifteen years, the science of weather prediction
in the United States has advanced significantly.   Weather forecasts
are now more accurate than ever before, and American meteorologists
have been able to enlarge the period covered by their forecasts.
Twenty years ago, meteorologists could not forecast the weather four
or five days ahead with significant skill.   Today they can do so.
A measure of our forecast improvement is that today we can forecast
the winds in the mid troposphere four days in advance with the same
accuracy that we could previously forecast two days in advance.
And forecasts at the surface have improved to a point where forecasts for four days are the equivalent in accuracy of previous three-
day forecasts.

     A number of technological factors have contributed to this
improvement.   One was the development of the electronic computer,
which permits us to analyze and process millions of bits of data at
high speed - and hence to use the physical laws governing atmospheric
state and motions as a basis of calculation.   Another was the

development of new communications links, which enable us to transmit
raw meteorological data with a rapidity to a national processing
center and to disseminate the meteorological products of the center
with equal rapidity.   And still another - and of equal importance
- was the development of the satellite, which made the inaccessible
areas of the world accessible to observation and information-gather-
ing.   For the continental United States, the satellite has been
particularly important for its ability to observe the atmosphere
over the adjacent oceans and over inaccessible inland areas like
the Rocky Mountains.   What does this advance in weather prediction
mean for society?   In the United States, improved forecasting is
of inestimable value to agriculture.   When to sow and when to reap,
when to fertilize and when to spray are vital questions, and better
forecasting means better crop management and protection.   During
the fruit-growing season in California, for example, an overnight
drop in temperature that brings it below the freezing point can
result in crop damage of $75 million.   But if we provide a reliable
forecast in ample time, growers can take action to protect their
fruit trees.   Today, because of the improvement in weather prediction, we can provide this forecast.   Let me add that the farmer or
crop owner is not the only beneficiary.   The world at large benefits.   If, as many predict, we are entering an era of chronic
world food shortages, crop management has worldwide significance.

Within the United States, the advance in weather forecasting
that new technology, including the satellite, has engendered is
equally of value to a wide variety of other activities.   Transportation benefits.   Improved forecasting permits us to route air
craft so that safety and comfort are enhanced, delays are avoided
where possible, and fuel consumption is minimized.   Construction
activities benefit.   Improved forecasting permits us to plan and
schedule construction work as environmental conditions demand.
Manufacturing activities benefit.   Improved forecasting permits
better planning of the manufacture of products whose sale and use
are sensitive to the weather.   And, simply as individuals, most
Americans benefit.   Improved forecasting permits us to enjoy
everyday life more by helping us to plan our recreational activities
- whether the activity be boating, bicycling, or taking the family
on an outdoor excursion.

3.2   I have stressed the satellite's contribution to the general
advance in weather forecasting because daily weather services are
perhaps the most important and most pervasive of the environmental
science services in everyday life.   But there are other meteorological benefits from the satellite, and they are uniquely due to the
satellite.   They come from the satellite's role as a platform for
weather surveillance.   My point is best illustrated by pointing
out that the day is now past when a hurricane spawned over the
south Atlantic or the Caribbean can strike the East or Gulf Coast

of the United States without prior detection.   Since the development of the environmental satellite, we have been able to watch
these storms as they form, build up their winds, and move.   They
sometimes take erratic turns, and we can track them as they do so.
These hurricanes are the single most destructive weather phenomenon
experienced by the United States.   A single storm can kill hundreds
of persons, wreak property damage in the millions, even billions, of
dollars, and cripple cities.   In 1972, for example, Hurricane Agnes
moved over the continental United States and killed one-hundred-and-
eighteen persons, caused property damage in excess of $3 billion,
and laid large areas prostrate.   But in every instance of a hurricane, we are able to give early warning to the areas that may lie
in its path so that people can take whatever protective measures
are necessary - to secure property as best they can, to take proper
shelter, and sometimes to evacuate a whole area.   Early warning
always means better protection.   In the case of Tropical Storm
Agnes, early warning resulted in the safe evacuation of eighty to
one-hundred-thousand persons from floods in Wilkes-Barre, Pennsyl-
vania.   And in the case of Hurricane Camille, which struck the
Gulf Coast in 1969, causing property damage of about $1 billion,
eighty-thousand persons were evacuated to safety because of timely
warnings.   In 1974, Hurricane Fifi struck Central America causing
great losses in Honduras.   The geosynchronous satellite permitted
excellent tracking and advance warnings.

        I have focused on hurricanes because our experience with the
detection and tracking of these storms is now a long one.   But, as
I indicated earlier, the geostationary satellite provides an excellent platform for the detection and monitoring of tornado-producing
thunderstorms.   These storms are small in scale; they come and go
in a matter of hours, and usually strike a relatively localized area,
often inflicting great damage.   The United States suffers some
eight-hundred tornadoes each year.   With conventional weather
facilities, it is extremely difficult to detect them in their early
stages and to follow them.   Radar plays an important role in
monitoring these storms.   So, too, does the high-resolution imagery
of the geostationary satellite.

        Let me illustrate.   In April 1974, the United States suffered
one of its most severe outbreaks of tornadoes.   In a period of less
than twenty-four hours, one-hundred-and-forty-eight tornadoes struck
an area comprising thirteen States.   More than three-hundred persons
were killed, and the damage to industrial property, farms and homes
was well in excess of half a billion dollars.   And yet, but for the
satellite, the toll would have been higher.   A national disaster
survey team concluded its report by saying that "radar and satellite
data were absolutely essential for effectively issuing severe weather
watches and warnings covering the large number of tornadoes spawned
in this outbreak".

Perhaps nothing dramatizes the uniqueness of the satellite more than the instances in which the satellite provided help where there was no other help.   Satellite pictures of polar ice have enabled us to find a way through the ice for vessels that would otherwise be trapped.   And satellite pictures have been used to assist ships at sea that were caught by ocean storms.   The satellite told us the position, the territorial coverage, and the movement of the storm, and this information was relayed to the ships in distress.   Obviously, what made the satellite so powerful a tool in these instances was that it was the only observing platform we had for these areas.   Without the satellite, we would have known nothing about the polar ice or the ocean storms.

What this means more generally is that we can now provide service where none was possible before.   We can provide oceanic weather information.   This new capability also opens up new ventures for man.   By providing weather services for Antarctica and for off-shore oil-drilling platforms, for example, we make it possible for men to undertake a long expedition to Antarctica or to engage in off-shore drilling - and to do so in comparative safety.

3.3   The satellite not only makes possible old services for new areas;   it also makes possible wholly new services.   Infra-red sensing from a satellite provides an excellent illustration.   Infra-red radiometers can scan the ocean surface and tell us its temperature, at least in cloud-free areas.   One of the things we can do with this information is map the Gulf Stream, which is of value to marine transportation.   Information about ocean surface temperatures is also of value to the fishing industry, for temperatures tell us where certain kinds of fish may congregate.

3.4   The advances in weather forecasting and in the creation of new environmental services are significant achievements, but the greatest advances still lie ahead.   They will come from a programme of basic research into the dynamic processes governing the geophysical system that comprises the oceans, the upper and lower atmosphere, and the solid earth.   The high-speed electronic computer and the earth-orbiting satellite have opened the door for this research.   Fifteen years ago, there was no way that the environmental scientist could deal with the global environment.   But the computer now makes it possible for the environmental scientist to simulate the physical processes of the geophysical environment mathematically and to experiment with his model.   And the satellite has opened up the entire globe to satisfy the scientists' passion for environmental data.

3.5   I said earlier in this paper that there is a long tradition of co-operation in weather matters.   It grew out of the mutual dependence of nations for weather information.   Weather is an international phenomenon.   It knows no borders, and nations have always

needed to know about their neighbours' weather for their own domestic
services.   And so the free exchange of weather information became
customary.   Nations participated in the exchange because it was in
their own best interest to do so, and each benefited from the
exchange.

4.   INTERNATIONAL   CO-OPERATION

We can trace the beginnings of international weather co-operation back to the middle of the seventeenth century, but the first
modern international co-operative effort was established by the
Brussels Maritime Congress in 1853.   The maritime nations agreed
to make regular observations of ocean weather from merchant vessels
and to exchange the observational data.   The venture was successful,
and twenty years later led to the formation of the International
Meteorological Organization, the first international body devoted
to the planning and co-ordination of observing networks and to the
standardization of meteorological instruments and techniques.   Its
membership was primarily European.   In 1951, it was superseded by
the World Meteorological Organization, a specialized agency of the
United Nations.   Its membership is global.

The United States developed and constructed its environmental
satellite system primarily to meet its own domestic needs.   It
wanted an observing platform for its adjacent oceans and its moun-
tainous regions in order to improve its regular weather forecasting.
And it wanted an observing platform to keep watch on severe storms.
At the same time the United States recognized that the satellite
would also provide it with the worldwide data essential for the
improvement of general weather forecasting (increasing their accuracy
and extending their time range).   The value to other nations was
clear.   In the tradition of international weather co-operation, the
United States has readily made the information gathered by its
environmental satellites accessible to others.

4.1   The service most widely used by other nations is automatic
picture transmission.   The present estimate is that there are now
approximately seven-hundred low-cost APT receiving sets in use
throughout the world capable of reading out the images transmitted
by the NOAA satellite as it passes overhead in its polar orbit.
Of these, approximately five-hundred are in non-American hands.
Of the one-hundred-and-forty countries and territories that are now
members of the World Meteorological Organization, one-hundred-and-
nineteen have at least one APT receiving set.   The Soviet Union has
many installations;   and when China opened its doors to the West, it
was discovered that it had over thirty sets in operation.

The United States does more than simply say to the world that
its satellite information is available for "read-out" by an APT set.

It also provides technical information on the satellite and its on-
board equipment to make the task easier.   And it provides assistance
in processing and interpreting what is read out.   This was particularly important, for example, when the United States switched from
video cameras to scanning radiometers because the new images required
some adjustment after they were received.   The NOAA satellite also
transmits directly the images scanned by its very-high resolution
radiometer.   Because of the high cost, however, only a handful of
countries have a receiving set capable of reading out high-resolution
picture transmission.   Finally, the NOAA satellite transmits directly
the temperature soundings taken of the atmosphere.   Even fewer countries are capable of receiving these transmissions.   Here the deter-
rent is the high cost of the data processing equipment.   However,
these data are processed in the United States and made available
throughout the world on weather communications circuits.

4.2   There is also WEFAX, the broadcast service provided by the GOES
satellites.   WEFAX, like APT, requires relatively low-cost receiving
equipment.   The service, of course, is limited geographically to the
area irradiated by the satellites.   But for this area, WEFAX is an
important supplement to APT.   I should add that all the direct-
transmission services I have described are rendered by the United
States at no cost to a recipient country other than the cost of
installing ground receiving equipment.   There is no charge for the
services themselves, nor does the United States demand a commitment
from a recipient country that it will in return make its environmental data freely available.

4.3   For many nations, the American environmental satellite system
has become an invaluable adjunct to their own environmental informa-
tion-gathering systems.   The satellite enables those that are adjacent to the oceans to get atmospheric information (which is vital to
their domestic weather forecasting) that they could never get before.
France and Australia are examples.   The satellite has also enabled
some nations to develop new or improved environmental services.
Iceland, Canada, Argentina and New Zealand, for example, use satellite data for the reconnaissance of ice and snow.

     The nations I have named are in the developed world;   they have
the technical competence to make the widest possible use of the
satellite.   But the satellite has also been of immeasurable help to
the developing countries.   For the most part, their weather services
are sufficiently developed so that they have been able to make large
strides forward through the use of satellite data.   In many of these
countries, conventional data are sparse because of a lack of observing stations or of communications facilities, and the satellite has
become a unique source of weather data for forecasting.   Here we
have an outstanding example of an instance in which the advanced
technology of a highly developed nation has brought very tangible

benefits directly and immediately to nations that are beginning to climb the ladder of technological development.

4.4 The satellite is also at the heart of two new and very significant efforts in international meteorological co-operation. One is the World Weather Watch, a comprehensive system for the observation of the global environment and for the rapid and efficient processing, analysis, and dissemination of worldwide weather information. The Watch is being coordinated by the World Meteorological Organization, and it will make the fruits of modern technology available to all nations. In addition to its polar orbiting and geostationary satellites, the system has three world meteorological centers – in Washington, Moscow and Melbourne – and they are supplemented by a number of regional centers. These centers employ the latest processing and analytic equipment and techniques; and they communicate their meteorological products globally or regionally through regular international communications links, which have become increasingly rapid over the past decade.

The United States regards the World Weather Program as an international co-operative venture of the highest importance. In June of this year, President Ford submitted to the Congress of the United States my government's plan for participation in the Program for the coming year. In his letter of transmittal, the President reiterated the importance the United States attaches to the Program: "People everywhere recognize that weather influences day-to-day activities. People are also mindful that weather, sometimes violent, breeds storms that take lives and destroy property. Coupled with these traditional concerns, there is now a new awareness of the cumulative effects of weather. The impact of climate and climatic fluctuations upon global energy, food and water resources poses a potential threat to the quality of life everywhere. The World Weather Program helps man cope with his atmosphere. We must continue to rely upon and to strengthen this vital international program as these atmospheric challenges – both old and new – confront us in the future".

4.5 The second international effort is what has come to be called the Global Atmospheric Research Program (GARP). The programme will extend over many years and embraces a series of regional and global experiments dealing with the general circulation of the atmosphere and its predictability. The purpose of the programme is to acquire a deep scientific understanding of this circulation to lay the basis for improved long-range weather forecasting and the development of our knowledge of climatic fluctuation. The programme will require extensive data-gathering, and some experiments have already been carried out on a regional basis. The most notable of these has been the Atlantic Tropical Experiment, the largest international scientific co-operative effort ever undertaken, in which the satellite played a key role.   Daily scientific missions to investigate cloud systems were decided on the basis on satellite photography. In 1978, the first GARP global experiment will be carried out, in which about ten satellites will be used.

With the evolution of satellite technology and with growing experience in the use of satellite data, other nations have come to regard the satellite as almost an integral part of their own domestic weather services.   The sense of mutual dependence that has traditionally characterized international weather co-operation has taken on a new dimension because of the growing dependence of nations on the satellite.   The Singapore weather service has gone so far as to state: "It is not exaggeration to say that we now treat meteorological satellite data as of greater importance in providing information on the existence and development of weather systems over our region than the routine surface synoptic charts". And quite recently, in June of this year, the Director-General of the Argentine Meteorological Service wrote me about the use the Service makes of the weather data supplied by the World Weather Center in Washington.   He concluded by saying: "I feel it is my duty to inform you that a great part of the forecasting done by the Regional Meteorological Center in Buenos Aires would not be possible without the meteorological satellite data furnished by NOAA".

4.6 Dependence on the satellite, of course, means dependence on the countries that have the technical and financial capability to deploy environmental satellite systems.   At the moment, there are two – the United States and the Soviet Union.   By 1977, they will be joined by Japan and by the European Space Agency (the newly created successor to the European Space Research Organization), which is a consortium of Western European nations.   Japan will place a geostationary environmental satellite in orbit over the Western Pacific, and the European Space Agency will place one in orbit to serve Europe (and at the same time, Africa).   In addition, about 1978, the Soviet Union will supplement its polar-orbiting satellites with a geostationary satellite.   It will be placed in orbit over the Indian Ocean.

As other nations have become increasingly aware of their dependence on satellite data, questions have naturally arisen about the kind of international system of environmental satellites that will evolve.   Nations that are dependent on satellite data want to be able to participate in the process of deciding how satellites should be designed, where and when they should be deployed, what kinds of equipment they should carry, and what functions they should perform. The precise contours of internationalization have not been explored. These questions have yet to be discussed in an international forum.

4.7 The issue poses a serious dilemma for the United States and for other satellite-launching nations.   The United States has placed

its environmental satellite system at the service of mankind and has
made the information gathered by the system accessible to all
nations.   It has made this commitment as part of the World Weather
Watch, a programme that was developed and agreed upon internationally.   The United States has consulted with other nations and
has given their requests fair consideration.   But in the end, it
cannot compromise the domestic services that its satellite system
renders to the American people.   The system is crucial to their
safety and well-being, and the United States cannot sacrifice this
ultimate interest.

Two examples will make my point.   When the United States substituted scanning radiometers for video cameras, a number of nations
complained.   Because the video camera was easier to read out with
an APT receiving set than was the radiometer, they wanted the video
camera retained.   Yet the radiometer yields more valuable data,
and that is why it was substituted for the video camera.   The
United States cannot delegate this kind of choice - which affects
the essential quality of its system - to an international body.

My second example is more critical.   It involves daily management of the American system.   When there is a threat of tornadoes
in the midwest portion of the United States, the GOES satellites
are used for limited scanning.   Ordinarily, they would scan the
same area once every thirty minutes; on a storm day, they scan
once every fifteen minutes.   But to do so, scanning must be confined to the Northern Hemisphere and South America cannot be covered.
Would international management permit this kind of treatment?
Probably not.   And yet the GOES satellites were designed by the
United States for just this purpose - to keep a close watch on
severe storms that threaten its own territory.   The United States
can ensure that this watch is kept only if it retains the management
of its own environmental satellite system.

4.8 We know from other international endeavours that certain kinds
of internationalization pose other problems.   One is the protrac-
tion of the decision-making process.   For an environmental satellite system, it will mean a considerable slowing of the pace of
technological development and innovation.   Another is the risk of
politicization.   My concern is that a group of nations may seek to
bar another nation from participating in the benefits of certain
types of internationalized systems purely for political reasons.
And that kind of action will surely weaken the tradition of international co-operation in the environmental sciences.

On the other hand, there is no way in which satellite-launching
nations can avoid the participation of the international community
in environmental satellite systems.   Indeed, this participation is
a necessity for future progress in the growth and improvement of the
environmental services throughout the world.

4.9   There are now two operational polar-orbiting satellite systems,
one American and the other Russian, and two American geostationary
satellites.   In a few years, there will be three additional geostationary environmental satellites - one Japanese, one European and
one Russian.   Taken all together, these satellites will cover the
globe continuously.   To date, meteorologists have been successful
in organizing internationally co-ordinated systems of various kinds
to meet the humanitarian needs of all nations.   Weather data are
simply not withheld except during time of war.   The United States
today routinely alerts all nations, both those with whom it has
normal relations and those with whom it does not, of impending
weather hazards detected by its satellites.   This tradition is so
ingrained that it will continue.

    What are the primary interests of the non-satellite nations in
environmental satellite systems?   There are several.   The various
national and regional satellite systems should form a co-ordinated
whole.   Moreover, the satellites must be capable of satisfying the
basic requirements of the international community for observational
data, and there should be an assurance that all nations will have
access to the information gathered by the satellites.   We have the
World Meteorological Organization as the forum through which we may
achieve these goals.

    Beyond all this, the non-satellite nations must develop the
capability, either domestically or on a regional basis, to make the
best use of satellite information.   In many instances, national
and regional facilities will have to be markedly improved;   and
they will have to be co-ordinated with those of other nations and
regions.   There will also have to be assistance of various kinds
by the developed world, particularly by the satellite nations.
The task is an international one.   It will be long and difficult,
but it can lead to very large benefits to all mankind.

THE IMPACT OF SPACE ASSISTED METEOROLOGY

Erik Tandberg, Director, Norconsult A/S, Norway

An edited summary of the discussion

1. ADDITIONAL NOTES ON INTERNATIONAL ACTIVITIES

The European Space Agency's METEOSAT meteorological satellite programme is optional. In other words, each member state has the choice of joining or not, and at the time of the Symposium eight out of ten members participated in the programme.

METEOSAT is one of the five satellites in the world experiment outlined by Robert M White in his paper. It is also the most advanced of the European Space Agency's application programmes. During the development phase the Space Agency has stressed good relations with the users through continuous dialogues. The Agency has explained what METEOSAT could do, has listened to what the users wanted and has made an effort to see how far it was possible to go to satisfy them. The first launch is scheduled for the middle of 1977, and in recent months it has been decided by the users to entrust the Space Agency with the operational part of the programme for the first period. The original plans did not include this possibility.

The European Space Agency also termed the METEOSAT programme an outstanding example of the way in which it is possible to co-operate internationally and meaningfully. The Agency has a system of collaboration with the Soviet Union, Japan and the United States, a system where the informality is unusual and yet very efficient. It paid tribute to the characteristic co-operation that has existed for many decades among meteorologists, a co-operation that the European Space Agency is profiting from to a considerable degree. Hopefully this spirit will also find its way to other fields in the years ahead.

111

2. SATELLITES AND AIR POLLUTION

Atmospheric pollution is indeed a complex problem. The pollutants vary in character and quantity, they may appear in different altitudes, and their presence will sometimes be a local, sometimes a global problem.

Carbon dioxide is a well-known pollutant, and its increase over the years is mainly due to the burning of fossil fuels. There is not much short-time variation on the global scale, however, and since today's satellites cannot measure gaseous constituents of the atmosphere with the sort of accuracy needed, they are not particularly valuable means of monitoring changes in the carbon dioxide level.

But there are other atmospheric constituents which could possibly be monitored by satellites, minor constituents whose variation is not so uniform on a global scale and which are linked with man-made sources. Some are connected with ozone photochemistry, which has become particularly important in recent years. Experiments on the NIMBUS-G will set out to measure in the stratosphere some of the gaseous constituents which are only present in a few parts per billion, for instance, nitrogen oxide and nitrogen dioxide. The mapping of carbon monoxide, one of the milder constituents from man-made sources, should also be feasible from satellites.

Local pollution is most often connected with atmospheric constituents near the ground or in the lower atmosphere, and is very difficult to monitor from space. Two reasons for this are the background provided by the earth's surface and the dust near the ground. A third reason is the presence of clouds and rain, two of the factors controlling the concentration and the dispersion of pollutants in the lower atmosphere - especially particulates, because these are easily washed out by the rain. Satellites have, of course, difficulties making measurements and observations under cloud and rainy conditions.

Given time and resources, instruments will probably be developed for quantitative identification of most gases in the atmosphere. Laser systems used with two satellites or with one satellite and several ground stations are certainly a possibility, but it is also conceivable that low altitude polar orbit satellites could carry special spectrometers and radiometers to identify most of the important gases. Not all of them, and not with the accuracy that many scientists would like to have, but sufficient to combine the measurements with those from conventional environmental satellites in geostationary and low polar orbits and from the earth's surface to develop a tractable global and regionally significant pollution monitoring system. Such a system would probably come about in the 1980's, and more likely in the later '80's than the earlier '80's.

3. CAN SATELLITES CONTRIBUTE TO CLIMATE ANALYSIS?

The difference in predicting changes of climate as opposed to predicting changes of weather is primarily one of time scale. This means that from the point of view of prediction there is additional information needed in order to predict climate, information not needed to predict day-to-day weather. There is no method at the moment - neither empirical nor numerical - that will predict changes of climate. To develop such a method it is necessary to monitor changes in the atmosphere, on the surface of the earth, and in the solar radiation to get an idea of what major processes produce changes of climate.

Most scientists believe that the method will involve the use of complex numerical models. Such models have been constructed and are in use in the United States, the United Kingdom and possibly one or two other countries. They are very complex, and demand a tremendous amount of computing time.

In the basic physical approach to weather forecasting, equations of physics are integrated for just a few days. In order to run models for climate, the equations have to be integrated for hundreds of days, or, if longer-term climatic changes are wanted, for much longer periods. This raises the question whether there are models which will remain stable for that length of time and whether the computers are big enough to undertake the enormous computing task. Models have now been integrated half an hour at a time up to a year or more, and it is possible from these models to produce the present day climate. Major features such as the monsoons, the main wind systems, the distribution of the rainfall, and the temperature can be simulated quite realistically. However, this is not the same as making a forecast, and the next thing to do is to use the models as an experimental test-bed to find what could change the climate. Parameters such as the solar input, the distribution of land and sea, the ice cover, the topography, the atmospheric constitution, the proportion of ozone, carbon dioxide, etc., can be varied, but the result may not be the right one. The models could still be far from complete, far from containing all the basic physics really needed for a climatic model. To some of the Symposium participants the influence of aerosols, atmospheric dust, the earth's magnetic field, the ozone layer and the release of energy here on earth represented the major unknowns. Others were of the opinion that the interaction between the ocean and the atmosphere was the biggest problem. Or, to put it another way, to predict the weather for a few days ahead it is not necessary to know too much about what is going on in the oceans as long as the ocean temperature is known. But because of the tremendous storage of heat in the ocean, and the fact that everything there occurs on a time scale about one-hundred times more slowly than in the atmosphere, changes in climate are

probably intimately tied to what happens in the ocean.   To really
understand climate changes it is important to establish models of
a joint ocean/atmosphere and predict what goes on in the ocean.

Changes in the climate will, of course, have a profound effect
upon several areas of human activity.   One is food production.   It
is a fact that the first half of this century has been one of the
warmest, most benign and most suitable periods for food production
since the 14th or 15th century.   Indications are that the earth has
recently been experiencing a cooling trend.   It is not known whether
this trend will continue and produce a cooler and more severe climate
than in the first half of the century.   The consequences of a con-
tinued cooling trend because of over-population and the other
environmental problems may be serious.   Can satellites contribute
to climate analysis?   Probably, in many critical ways.

Whatever climate models are developed we will still need to
know about the initial and boundary conditions, to predict the
fluctuations of the general circulation of the atmosphere.   Therefore initial conditions that are being observed by today's satellite
systems, such as temperature and wind, will continue to be very
important.   Satellite techniques will be used in conjunction with
other methods to learn more about the processes that have a bearing
on climatic trends and on the boundary conditions that are so important in determining climate changes.   Changes in rainfall, trends
in the global cloud cover, and ocean temperature changes are
examples.   Ultimately climate models must take into account ocean
circulation, ice cover, the planetary radiation budget, etc.   The
monitoring of these and other parameters could be done by satellites.
Satellites should be able eventually to study the size, distribution
and reflective indices of aerosols in the atmosphere, the amount of
ozone, changes in the earth's magnetic field, etc.   One important
parameter that will be very difficult to study from space is the
deep ocean circulation.   There is a possibility, however, that
measurements can be made by a combination of special, deep sensing
buoys and satellites.

4. NATURAL DISASTER WARNINGS FROM SATELLITES

Environmental satellites can assist in providing warnings about
natural disasters that stem from atmospheric and oceanic events.
Present satellites are used on a worldwide basis for providing
critical information on weather hazards such as hurricanes and
severe storms.   Satellites may be able in the future to assist in
warning of earthquakes.   Warnings are effective only if the warnings can reach the people who must take action.   We are presently
able to transfer information from the observations to those in a
position to make predictions.   It is more difficult in most parts

of the world to get the warnings to people in exposed areas in time
and to really make them heed the warnings.   There are experiences
even in the United States where warnings have been transmitted
but the people are so used to their daily conveniences and their
relatively sheltered life that they do not pay enough attention to
them.

There are also instances where all the people cannot be reached
fast enough, for instance, campers in a valley exposed to a flash
flood.   Here devices which can get to people very quickly would be
useful.   In the United States a preliminary study has been done on
a disaster warning satellite which by direct broadcast should be
able to reach all the people who need the warning.

It is also possible that valuable time can be saved in the
transfer of information to users by placing trained meteorologists
in space.   Valuable experience can be gathered with the Soviet
SALYUT space stations and the European/American SPACELAB/shuttle
programme so that the value of a manned early warning system can
be assessed when more permanent space stations are operative, hopefully even in geostationary orbits.

5. CAN SATELLITES IMPROVE WEATHER FORECASTING FOR
AIR TRAFFIC, MARINE TRANSPORTATION AND OFFSHORE OPERATIONS?

As far as scheduled aviation is concerned, the weather forecasting of winds and temperature for the en route operations is now
to a large extent automatic, and seems to meet the needs of airlines
under normal conditions.   Robert M White indicated that in the
United States severe weather en route is a critical problem where
aviation could be assisted by information from geostationary satellites.   Other participants wanted more data on jet stream and clear
air turbulence location.   However, it was not made clear to what
extent satellites could be of assistance here.

Geostationary satellites can help to improve terminal weather
forecasts, while it will not be possible for such satellites to
obtain information on ceilings and runway visual ranges, it should
be possible for them to provide key data on severe weather.

In marine transportation several countries today offer, on a
commercial basis, a ship routing system where the individual vessel
is given its best route either for maximum speed or for minimum
damage.   This routing system is based on wind field and sea state
forecasts, but is good for only forty-eight to seventy-two hours.
If forecasts could be given for seven days ahead, the optimum route
would be indicated in the very early stages of long trips.   This
would be an improvement, and it is very likely that satellite systems will play an important role in the collection of data for such

forecasts.   Another improvement also tied to the use of satellites
would be the ability to define ocean current systems better.

     Weather forecasting for offshore operations has always been
important because of the capital investments and the environmental
aspects involved.   In the North Sea, the rapidly increasing activity
has meant more involvement from and more challenging tasks for the
meteorological services in countries like the United Kingdom and
Norway.   A few years ago the Meteorological Office in the United
Kingdom did North Sea forecasting for a few thousand pounds a year;
now the work is getting up towards half a million pounds a year.
The Meteorological Office is currently forecasting for about forty-
five different locations, which is demanding because it is a question of predicting weather and sea state conditions at points only
in one of the most violent stretches of ocean in the world.   The
tow-out of large, very expensive rigs often requires a high proba-
bility forecast of four to five days of absolutely calm weather,
which does not make the task any easier.   Operational meteorological
satellites will continue to be one of the information sources for the
forecasts.   In addition, new sensors and new satellite systems,
such as NASA's SEASAT, will facilitate a closer monitoring of the
sea state conditions.   Future sea state information could also con-
ceivably be collected by over the horizon radar systems.

   6. WILL THE EXPENDITURES INVOLVED IN EXTENDING WEATHER
       FORECASTS TO THREE WEEKS BE WORTH THE EFFORT?

     It is, of course, very difficult to put a cash value on many
of the benefits that flow from improved weather forecasting.   On
the other hand, serious attempts have been made to do this, and a
fairly detailed study was done before it was decided to set up the
European Centre for Medium-Range Forecasts.   Here they are not
even talking about three weeks, as the terms of reference are from
four to ten days, but if the Centre could produce a reasonably reliable forecast for about one week ahead, the benefits have been cal-
culated to outweigh the costs by a factor of something like twenty
to one.

     Farming, at least in Europe, is an example of a weather sensi-
tive activity that would benefit considerably from a longer range
forecasting than available today.   The reason is obvious - tactical
decisions on ploughing, harvesting, crop spraying, haymaking, etc.,
could be made with confidence for instance a week in advance, and
a work schedule planned accordingly.   However good a forecast is
for one or two days it does not really give the farmer enough time
to change his work schedule.   The situation is equivalent in other
areas of activity, and it has been estimated by the building industry
in the United Kingdom that the losses due to bad weather are in the

order of two-hundred-million pounds a year.   Much of this is because
the managers and the supervisors cannot reschedule the work very
easily.   If a really good forecast was available a week in advance,
it has been estimated that the losses could be reduced by something
like ten to twenty per cent.   The shipping and the power supply industries are other examples, and it was stated at the Symposium
that the Central Electricity Generating Board in the United Kingdom
would like to have forecasts of hour by hour temperatures accurate
to one degree Centigrade for a week or more ahead so that they could
estimate the load and schedule the plants accordingly.   Of course
it will take time to develop a system capable of forecasting temperatures with that accuracy.

It was the consensus of the Symposium participants that the
expenditures involved in extending meteorological forecasts were
really worth the effort for weather-sensitive industries (with the
possible exception of tourism).   The United States is making sizable
investments to develop reliable forecasts a week ahead, a goal that
may be achieved within the next decade.   Satellites will certainly
play a role in these future forecasting systems, but will be more
advanced and specially suited than the present operational versions
which are designed to solve more immediate weather forecasting problems.

To develop reliable, three weeks weather forecasts is considerably more difficult.   In the opinion of several participants, one
of the reasons for this is the inability of the available computers
to handle the full amount of necessary equations.

PARTICIPANTS

ALFVÉN, HANNES — Professor, Royal Institute of Technology, Stockholm

BEAN, ALAN — Astronaut, U S A (Apollo 12, Commander Skylab 2)

BIGNIER, M — Director-General, Centre National d'Etudes Spatiales, Paris

BILLINGTON, R M — Consultant, External Telecommunications Executive, London

BLACK, J N — Principal, Bedford College, London

BODECHTEL, J — Professor, Zentralstelle für Geo-Photogrammetrie und Fernerkundung, München

BOLLE, H J — Professor, Meteorologisches Institut der Universität München

CHARYK, JOSEPH V — President, Communications Satellite Corporation, Washington D C

CLARKE, ARTHUR C — Sri Lanka

DAHL, HELMER — Director of Research, Chr Michelsens Institut, Bergen (Norway)

GIBSON, R — Director-General, ESRO, Neuilly sur Seine, France

HAEFNER, H — Professor, Institut für Geographie, Zürich Universität, Zürich

HOLBERG, K — Director of Research, Norway

HOPPE, G — Professor, Naturgeografiska Institutionen, Stockholm

HOUGHTON, J T — Department of Atmospheric Physics, Oxford University, Oxford

LIED, FINN — Director, Chairman, Executive Board Norwegian Council for Scientific and Industrial Research, Norway

LOCHER, FRITZ — Director-General, Post, Telephon und Telegraphenbetriebe, Bern

LÜST, REIMAR — Professor, President, Max-Planck-Gesellschaft, München

MASON, B J — Director-General, Meteorological Office, Bracknell, U K

MOE, JOHANNES — Rector, Technical University of Norway, Trondheim

NORDBERG, WILLIAM — Director of Application, NASA, Goddard Space Flight Center, Greenbelt, Maryland

OMHOLT, ANDERS — Chairman, Space Committee, Norwegian Council for Scientific and Industrial Research, Norway

ORTNER, J — Director, Österreichische Gesellschaft für Weltraumfragen, Wien

PETERS, B — Professor, Danish Space Research Institute, Lyngby

SJALINOV, VALERIJ P — Scientific Secretary, Institute for Space Research, Soviet Academy of Science, Moscow

TANDBERG, ERIK — Director, Norconsult A/S, Norway

WHITE, ROBERT M — Administrator, National Oceanic and Atmospheric Administration, Rockville, Maryland

INDEX